I0894393

Key Concepts STEM

SCIENCE TECHNOLOGY ENGINEERING MATH

ENGINEERING AND TECHNOLOGY

POWER AND ENERGY

General Editor: Anne O'Daly

Brown Bear Books

Published by Brown Bear Books Ltd
4877 N. Circulo Bujia, Tucson, AZ 85718
USA
and
Studio G14, Regent Studios, 1 Thane Villas,
London N7 7PH, UK

© 2026 Brown Bear Books Ltd

Cataloging-in-Publication data available upon request.

All rights reserved. This book is protected by copyright. No part of it may be reproduced, stored in a retrieval system, or transmitted in any form or by any means, without the prior permission in writing of the Publisher, nor be otherwise circulated in any form of binding or cover other than that in which it is published and without a similar condition including this condition being imposed on the subsequent publisher.

ISBN 978-1-83572-065-3 (ALB)
ISBN 978-1-83572-071-4 (paperback)
ISBN 978-1-83572-077-6 (ebook)

Children's Publisher: Anne O'Daly
Design Manager: Keith Davis
Picture Manager: Sophie Mortimer
Consultant: Chris Woodford

Manufactured in the United States of America
CPSIA compliance information: Batch#AG/5664

Picture Credits
Front Cover: Shutterstock: Fokke Baarssen. **Interior: iStock:** alex 36b, Jer Barber 38t, baronozdemir 17t, danilove 38b, doomu 46b, Kadmy 14, Natalia Lagutkina 21b, mbbirdy 15b morkeman 18t, Jennifer Sanerkin 20t, Thinkstock 4b, 54, valigursky 23t, VanderWolf Images 19b, Sasa Vetkovic 21t, yongyuan 16, Zargon Design 6t; **Library of Congress:** 24bl, Charles Parsons 31; **Photos.com:** 6bl, 6br, 7, 8t, 10b, 12t, 18b, 27, 30, 41, 42tl; **Public Domain:** 37t, 53t, Augustin Privat Deshanel/Laplante 40t, William G Jackman 12b, Walter Larden 40inset, Newton Henry Black, Harvy Nathaniel Davis 28r, University of Pennsylvania/Edgar Fahs Smith Collection 43bl, Scottish National Gallery 28l, Chris Sunde 15t, TRUMPF 24r, Alfred John West/Our Navy 34b; **Shutterstock:** 10t, Arkstart 24br, Carlos Barquerro 50b, Belish 4t, Brude Alan Bennet 26, Bildagentur Zoonar GmbH 48, 61b, Bohbeh Creations 59, chippix 32, 61t, Hywit Dimyadi 35, dzmitrock 44b, Filip Fuxa 1, Tom Grundy 58b, Mykola Ivashchenko 22, Tim Jenner 8b, Phakorn Kasikij 57, KTS Design 43br, Kvsan 46t, Lornat 53b, John Mackintosh 34t, misha 11t, 60, R Moore 49, Gianluca D. Muscelli 58-59t, Norbor 52, PennyJack Creative 56, Vladimir Popovic 40b, rtem 36t, SocoXbreed 25, StacyStauffSmith Photos 20b, Bernhard Staehli 23b, Lisa Strachan 50t, Michael D. Vonka 58l; **USGS:** Doug Duncan 19r.

All other artwork and photography © Brown Bear Books Ltd

Brown Bear Books has made every attempt to contact the copyright holder. If you have any information about omissions please contact licensing@brownbearbooks.co.uk.

STEM

Key Concepts in STEM: Engineering and Technology describes the scientific principles and feats of engineering that have shaped the world we live in. Since earliest times, people have created tools and machines to provide shelter, warmth, security, and protection from disease. This book looks at the inventions that transformed the production of heat, light, and motion, including how water, fossil fuels, and electricity were harnessed to fuel the development of the modern world. It also looks at renewable sources of energy. Explanatory diagrams and informative photographs explain how STEM (science, technology, engineering, and math) has transformed our lives and continues to influence our world today.

Publisher's note to educators and parents: Our editors have carefully reviewed the web sites included at the end of this book to ensure that they are suitable for students. Many web sites change frequently, however, and we cannot guarantee that a site's future contents will continue to meet our high standards of quality and educational value. Be advised that students should be closely supervised whenever they access the Internet.

Contents

MAKING MACHINES

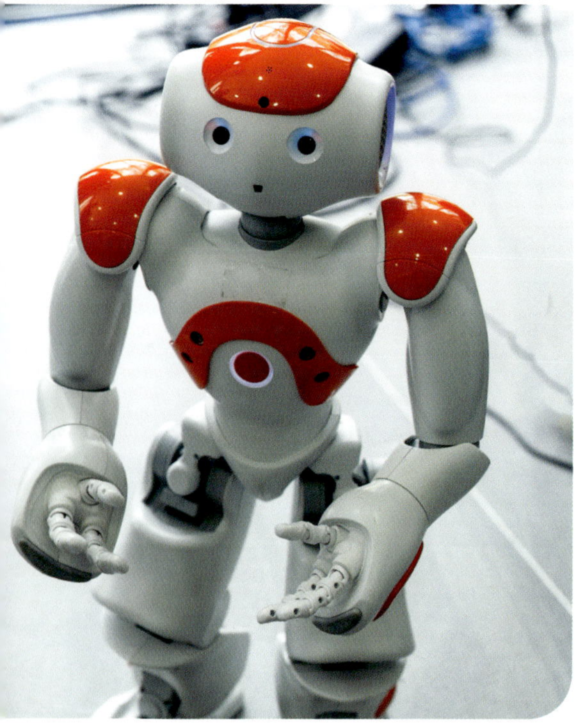

A humanoid robot is designed to copy the human body, with a head, torso, arms, and legs. Humanoid robots do not get tired or bored when working. Due to its friendly appearance, NAO (pictured left) is used in schools and healthcare settings.

The human body is perhaps the greatest and most versatile machine of all. Our skeleton works like an elaborate collection of levers and pulleys that can lift heavy objects and move at different speeds in a way that defies even the most advanced robots.

Despite its versatility, the human body has its limits, which cap how fast we can run, how much we can lift, and how much work we can do before we tire. For all the wonder of the human body the development of modern society has relied heavily on artificial machines to aid or replace muscle power.

Strictly speaking, a machine does not have to be a large, noisy contraption made of engines, gears, and wheels. It can be any device that converts a small force into a larger one. Simple machines include crowbars, scissors, wedges, and screws. The first machine ever invented was the hand axe, a fist-sized tool that was used by Stone Age hunters.

Raising Water

Agriculture has always relied on effective irrigation, and some of the first recognizable machines were devices for moving water from one place to another. The shadoof (also called the swape) enables a human operator to raise a heavy bucket of water several feet high using a long lever with a counterweight at one end. It is shown in images from around 2500 BCE that were made by the Akkadian civilization (adjoining the Sumerian region), and it has been used in Egypt since around 2000 BCE. More sophisticated water-raising

WEDGES

Wedges are simple machines that can open up or cut through hard materials. The thin part of the wedge is inserted into or under an object using force (and usually speed) applied from the blunt end. The wedge multiplies the force to open up or lift the hard material. Axes are wedges, and hand axes were some of the earliest human stone tools.

Edged weapons, such as this arrowhead, are wedges. The oldest human-made wedges are two million years old.

SIMPLE MACHINES

Machines transform forces into motion and back again. Five simple machines are the lever, wheel and axle, pulley, ramp, and screw.

1 A lever is a rigid bar pivoted around a point called the fulcrum. The lever transforms a small force (the effort) into a larger force (the load). Devices that use this principle include wheelbarrows, scissors, see-saws, and hammers.
2 A wheel and axle unit is effectively a lever with the axle acting as the fulcrum. If you turn a bicycle wheel by spinning the axle at the center, the outside of the wheel turns much more quickly but with less force.
3 A pulley uses a set of wheels and ropes so you can lift loads with less effort, although you have to pull the ropes a longer distance than the load itself rises up.
4 It is much easier to push a heavy cart up a ramp (wedge, or inclined plane) than it is to lift it vertically to the same height. Less immediate effort is needed, but that effort must be applied for a longer amount of time.
5 A screw, such as might be used in a car jack, is like an inclined plane wrapped around itself. The handle of the jack is a lever.

1. SIMPLE LEVERS

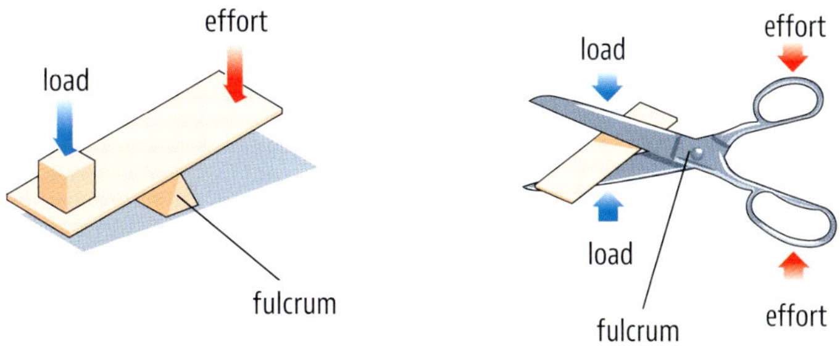

2. WHEEL AND AXLE

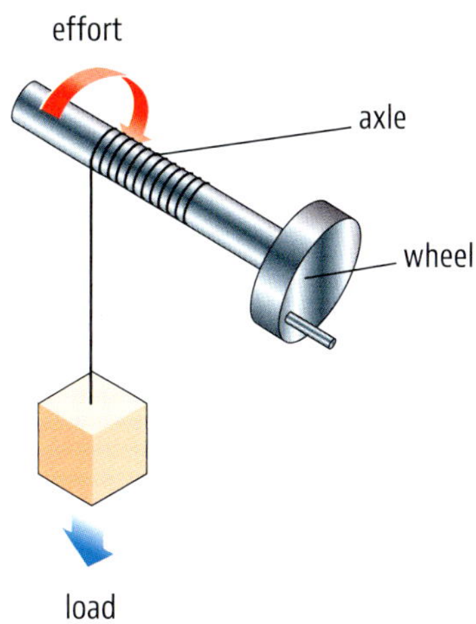

3. PULLEY

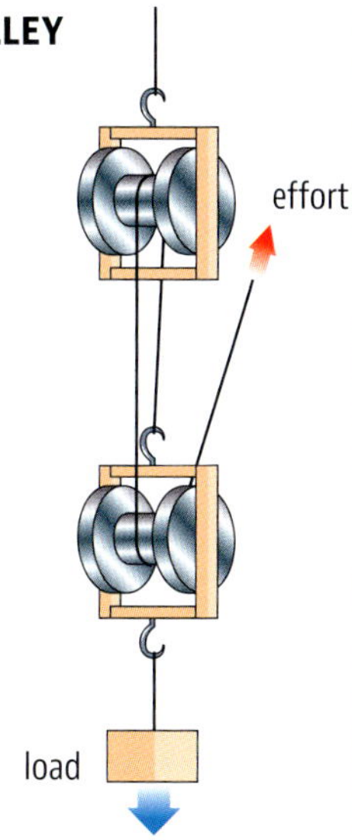

4. INCLINED PLANE

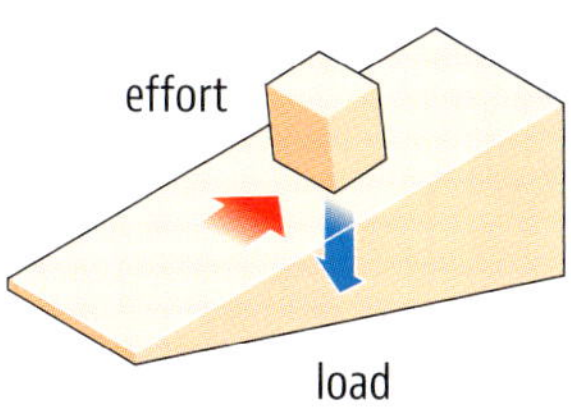

5. SCREW

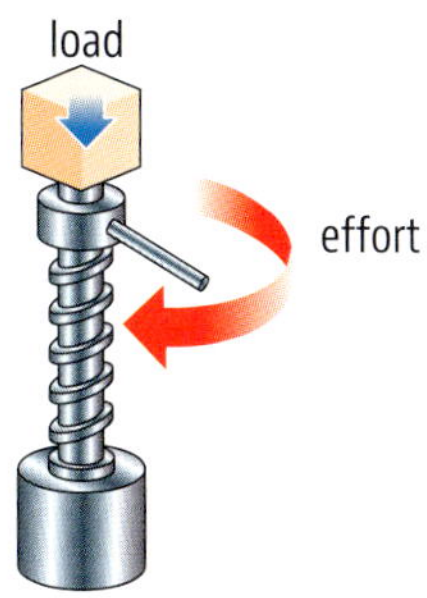

The mechanical advantage of a machine is a measure of how useful it is. A pulley with four ropes has a mechanical advantage of four, which means it could be used to lift a load of 20 lbs (9 kg) with a force equivalent to just 5 lbs (2.2 kg).

machines include the rotating auger screw, which is thought to have been invented by the ancient Greek mathematician Archimedes (c. 287–212 BCE) in the third century BCE. One version consists of a pipe leaning at an angle of 45 degrees with its lower end in water and a large screw wound around the inside. As the pipe is rotated, the screw lifts the water to the top, where it is collected. This type of a water pump is still used today.

PERPETUAL MOTION MACHINES

For many years, engineers have tried to produce perpetual motion machines that go on working forever without needing a fresh supply of energy. Some pumped water to a high point and then let it fall to drive the pump. However, these machines never worked. Energy is always lost as a machine works, so a resupply of energy is eventually needed to keep the mechanism working.

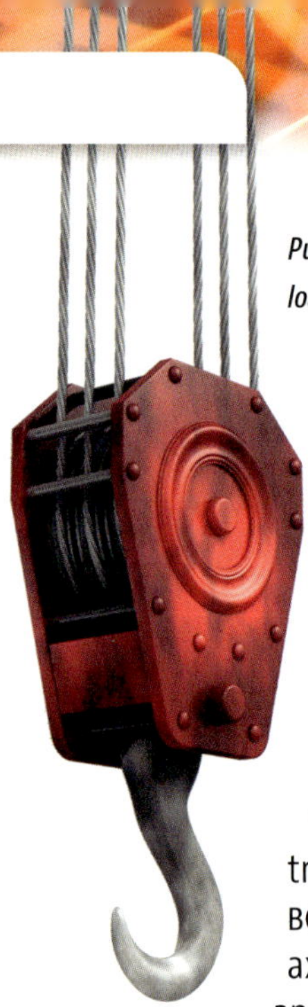

Pulleys use small forces to raise heavy loads. Ropes run around several wheels to amplify the force.

Rotary Motion

Potter's wheels were used to make pottery around the Turkish peninsula as early as 4250 BCE, but wheels were not used for transport until about 3,500 BCE. Images of wheels and axles have been found on ancient Sumerian clay tablets dating to that period.

It is thought that gear wheels emerged from ancient Greece. The philosopher and mathematician Aristotle (384–322 BCE) recorded the use of gears without teeth, and about 200 years later, the inventor Ctesibius of Alexandria (c. 270 BCE) developed a water clock that used gears. They were an immensely important invention and developed in two different directions. Large wooden gears became essential components in machines for transforming rotational motion in different ways, such as in water mills and windmills.

An old-fashioned illustration plate shows the workings of an Archimedes' screw. As you turn the handle (right), water rises up the inner screw.

SHADOOF

The shadoof is an ancient lever system used to raise water. It has a bucket suspended from a long arm that is pivoted near the other end. On the other side of the pivot, a heavy weight levers the bucket upward. The operator pulls the bucket down into the water to fill it. The counterweight automatically lifts the filled bucket out of the water.

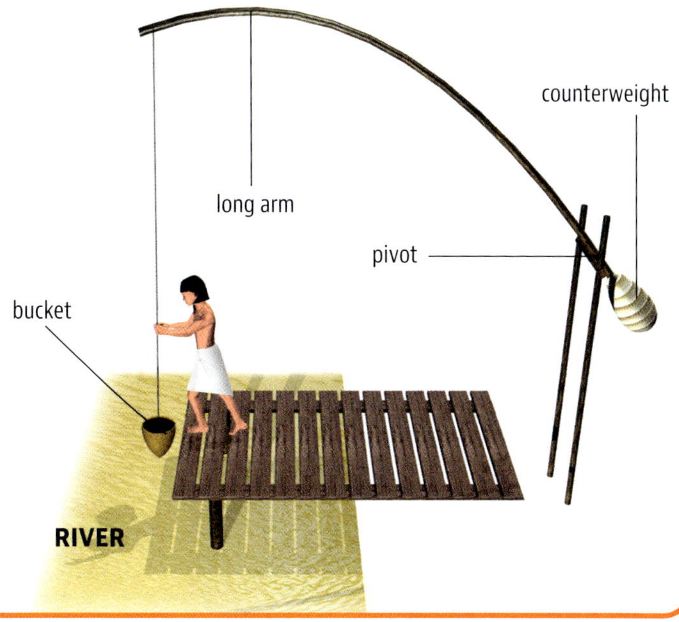

Shadoofs were used in early civilizations in the Middle East for taking water from rivers.

GREEK INVENTORS

Greek inventors were the original mechanics; the word machine comes from the Greek word *mechane*, which loosely translates as "device." The Greeks pioneered the making of mechanical toys called *automata* (giving us the word "automatic").

Although automata were designed for amusement, their ingenious mechanisms inspired many later inventions. Hero of Alexandria (in c. 62 CE), suggested a mechanical blacksmith's workshop (right), where water-powered automata hammered metals. He also developed numerous devices, including one for automatically opening the doors of a temple using hot air.

The power driving this potter's wheel comes from a drive belt moving around a spinning flywheel.

The Analytical Engine, Charles Babbage's mechanical computer, had thousands of interconnected metal cogs.

THE ANTIKYTHERA MECHANISM

In 1902, archaeologists found a corroded lump of bronze in a sunken ship near the Greek island of Antikythera. On closer inspection this turned out to be the remains of a complex gearing mechanism apparently dating from around 100 BCE. The surviving fragment contains more than 30 carefully filed gear wheels. It astonished and baffled its discoverers because it was as complex as clocks dating from the 17th century. Professor Derek de Solla Price of Yale University studied the device and believes it was an early mechanical calendar.

Centuries after the development of wooden gears, finely machined metal gears were developed for use in clocks and scientific instruments. In many ways, gears were the equivalent of today's electronic components. They were used in precision chronometers (clocks used for navigation on board ships), invented by British clockmaker John Harrison (1693–1776), and also in the early mechanical computers developed by British mathematician Charles Babbage (1791–1871).

Powering Machines

It was not until the Roman Empire (27 BCE to 476 CE) that machines started to make full use of rotating parts, such as treadmills and waterwheels. Although horizontal waterwheels were probably developed in ancient Greece, efficient, vertically mounted waterwheels were described by the first-century Roman architect and engineer Vitruvius only in 27 CE. The Vitruvian wheel was used to mill corn using crude wooden

GEARS

A gear is a wheel with teeth (sprockets) cut into its outer edge. These fit together (mesh) with the teeth of a second, similar wheel. As they rotate, one wheel moves clockwise and one counterclockwise (**1**). If both wheels need to move in the same direction, a smaller wheel called a central idler can be placed between them (**2**).

Gears can be used either to change the speed of rotation, or the torque (twisting force). If both gear wheels have the same number of teeth, they turn at the same speed with equal force, or torque. If one wheel has twice as many teeth as the other, it turns at half the speed but with twice as much force. If the larger wheel drives the smaller wheel (**3**), the speed is doubled and the force is halved. If the smaller wheel drives the larger wheel (**4**), the speed is halved and the force is doubled.

In cars, gears are used in both these ways. On flat roads, a gear is a speed multiplier to make the wheels turn quicker than the engine. Traveling uphill, gears are used as force multipliers. The wheels turn less quickly than the engine, but with greater force to pull the vehicle up the slope.

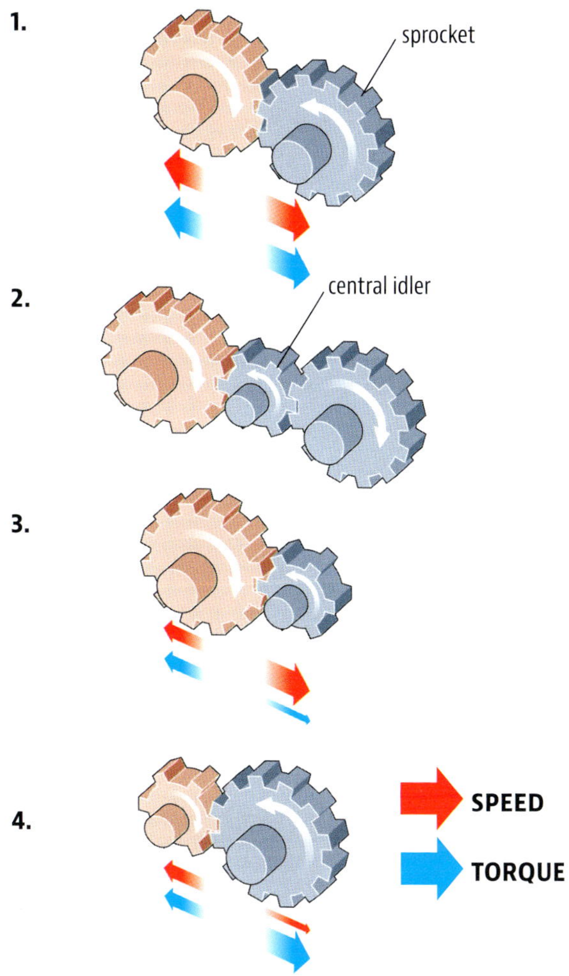

LEONARDO DA VINCI'S MACHINES

Leonardo da Vinci (1452–1519) is one of the most celebrated inventors in history. Although best known as a painter, he was also an accomplished architect, engineer, philosopher, and scientist. Da Vinci's notebooks contain drawings (right) of many original and ingenious machines, including water-raising devices, spinning machines, pumps, and machines of war. But this great Italian inventor is perhaps best known for his designs for intricate flying machines that resemble the modern-day helicopter, but with flapping wings. Leonardo's designs for flying machines were the result of his exhaustive studies of bird flight. He once wrote: "A bird is an instrument working according to mathematical law, which instrument is within the capacity of man to reproduce with all its movements." However, it was only when engineers realized that humans could never imitate birds that flight was finally achieved by the Wright brothers some 400 years later.

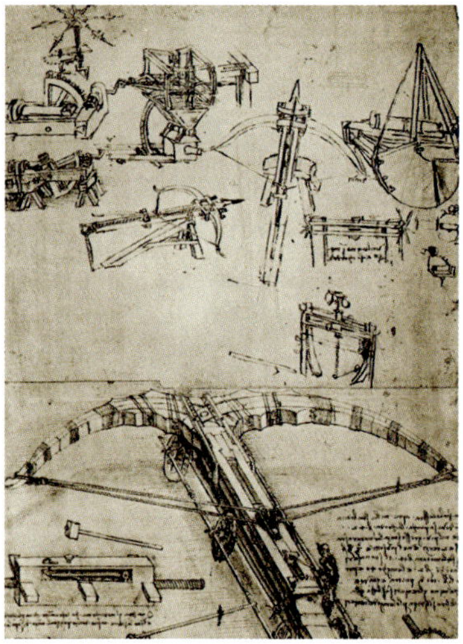

gears to drive heavy millstones. Later, water mills were used to power a variety of different machines. From the first century the Chinese were using water mills to make paper, and water-powered mechanical hammers were in use 200 years later. The use of water power spread rapidly after the end of the Roman Empire. By 1086 CE, when the Domesday Book (a comprehensive survey of English landholdings) was completed, no fewer than 5,600 water mills had been built in England.

Converting Power

By the end of the Roman Empire machines were using levers, wheels, pulleys, winches, and gears and were being driven by animals, humans, and water. From this time onward, machines became more complex and more refined. During the Middle Ages in Europe, machines began to evolve into the complex collections of wheels, gears, and levers that are familiar today. The key to this development was an understanding of how to

TREADMILLS

Romans powered their machines using treadmills (which were almost like giant mouse wheels) that were driven by humans to power corn mills, olive presses, water-raising devices, and early cranes. Roman treadmills were powered by slaves. In the 19th century, such treadmills were often powered by prisoners (below) and homeless people.

WINDMILLS

The windmill illustrated here was designed to mill grain. Other windmills have been designed to pump water out of fields or to generate electricity.

This grain mill has four large, wooden sails arranged in a cross. Attached to the center of this cross shape inside the windmill is the windshaft—a sturdy wooden axle that supports the sails and turns with them. The sails are of a type known as spring sails—they have small wooden shutters that blow open when the wind is strong, preventing the sails from bending and spinning around too quickly.

Earlier sails were covered in cloth that the miller had to roll up to slow down the sails. As the wind blows, the sails rotate, turning the brake wheel, which in turn rotates a series of gears, starting with the wallower. The wallower turns the great spur wheel, which then turns the gear wheels that move the millstones on either side.

The millstones are large, circular stones with a hole in the middle. Grain poured into the hole is crushed under the heavy wheel, becoming flour. Some windmills have a central post that allows the whole mill to rotate so that it is always facing the wind.

This old windmill in the Czech Republic was built to grind grain into flour.

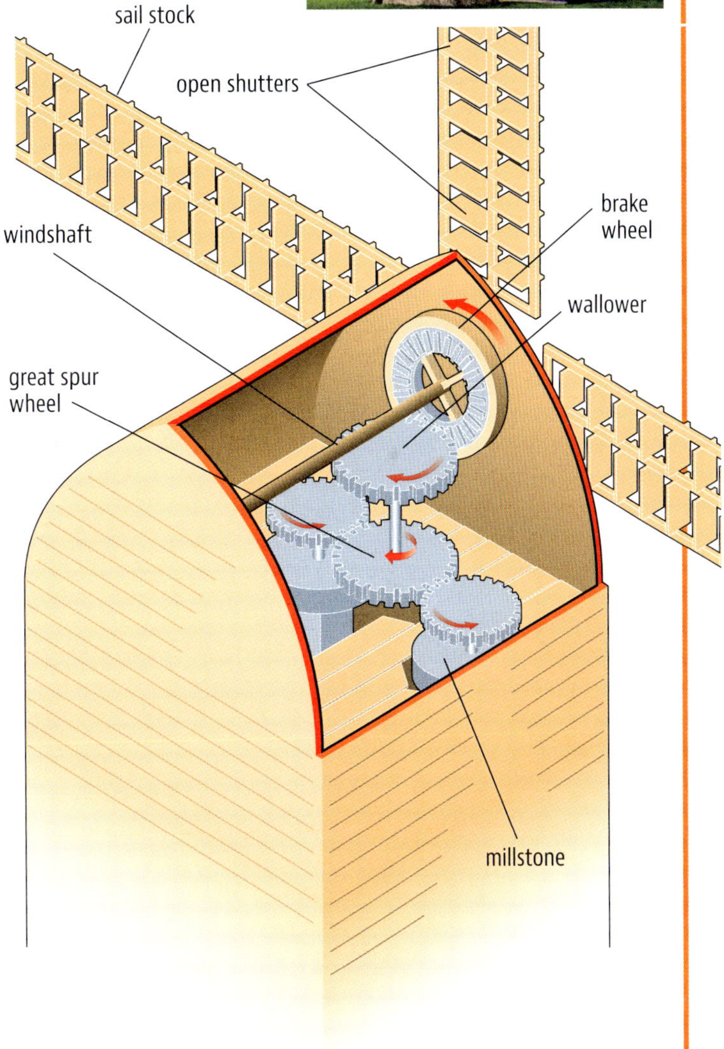

Windmills were invented in Persia in about 600 CE and spread to Europe in the 12th century. They enabled humans to harness the power of wind on land.

Just a handful of textile workers were needed to operate the steam-powered "spinning mules" invented in 1779 by Englishman Samuel Crompton (1753–1827).

transform power from a single source in various ways, and particularly how to change rotational motion into reciprocating (up-and-down) motion—and back again. Two mechanical inventions were particularly important for this process: the cam and the crank.

The Industrial Revolution

The Industrial Revolution was more of an industrial evolution—a gradual transition from the age of hand tools to societies that ran on machines. The Industrial Revolution began in Britain in around 1760, and has been spreading around the world ever since.

Machines changed in two basic ways during the Industrial explosion. First, a series of innovations in textile manufacture in Britain led to much more highly automated machines that enabled a single worker to produce a much greater output. Second, there was a revolution in the way in which machines were powered. Until this time, large-scale machines had been powered by water, and were built

OLIVER EVANS AND THE PRODUCTION LINE

The American Oliver Evans (1755–1819) was a pioneer of mechanization, where machines do the jobs once done by people. At the age of 25, Evans joined his two brothers at their flour mill in Philadelphia. He transformed the mill with water-powered machines, including bucket elevators and conveyors for raising the grain, auger screws to move it along horizontally, and automatic rakes to smooth it out. The beauty of the Evans mill was that each machine passed something on automatically to the next machine in the sequence without any need for human intervention. Grain was fed in at one end, and flour emerged at the other. This, the first ever production line, reduced the cost of the operation by as much as 50 percent.

CAMS AND CRANKS

The cam was first described in ancient Greece by the great inventor Hero of Alexandria. The device often takes the form of an egg-shaped wheel fastened to an axle on which one end of a rod rests. As the axle rotates, the egg-shaped cam raises and lowers the rod, thus converting the rotational motion of the axle into up-and-down motion in the rod. In automobile engines cams are often used to open and close the valves that let air and fuel into the cylinders.

The crank was first used in ancient China around 100 CE. It is effectively a handle bent out of the central axle of a wheel. A later modification, the crank and connecting rod, was made in 1430. Adding a rod to the crank that connects it to machinery allowed up-and-down motion to be converted into rotary motion, or vice versa. The crank can be used to convert the motion of a piston in a steam engine, for example, into the rotary motion of wheels.

In a four-stroke automobile engine a connecting rod links the pistons inside the cylinders to the crankshaft, which transmits (via the transmission) the up-and-down motion of the pistons to the rotating wheels that drive the vehicle forward or backward.

CAMS

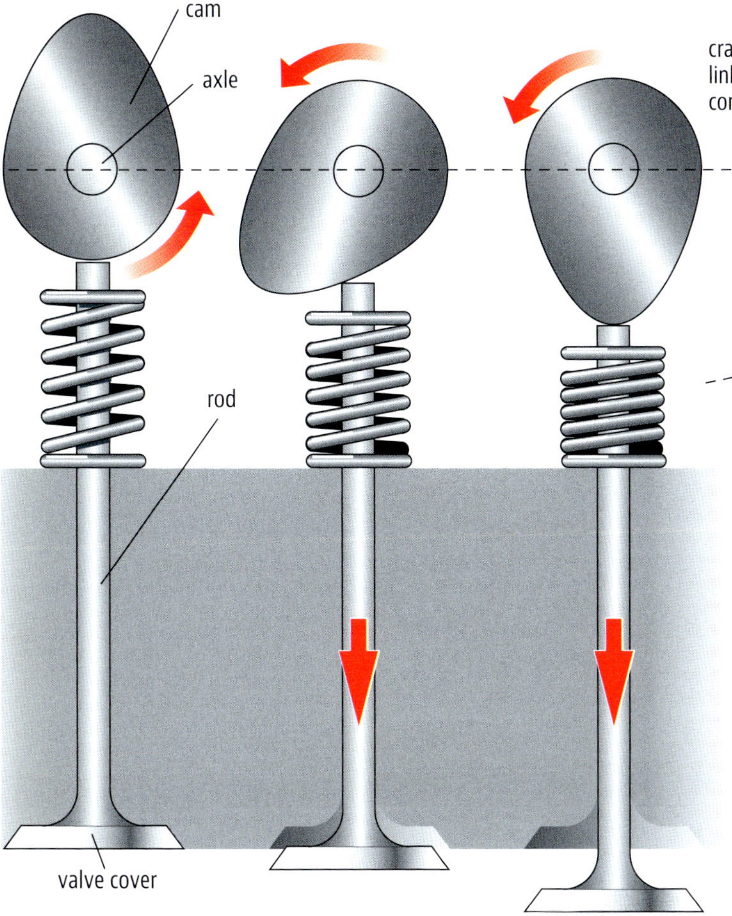

CRANK

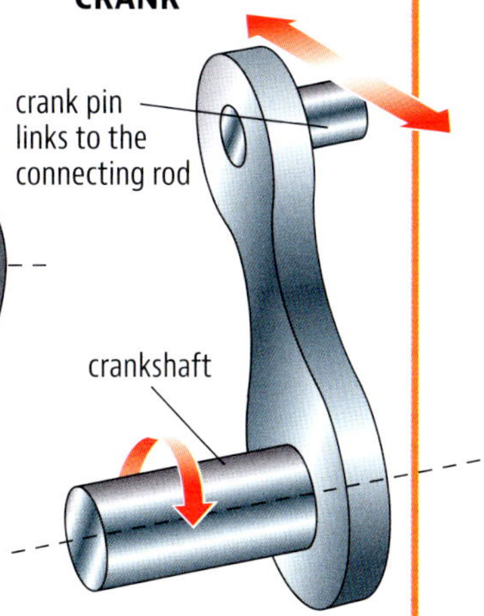

↑ *A crank is used to convert an up-and-down motion into the rotation of the crankshaft.*

← *Cams are used to convert rotation into an up-and-down motion.*

An earthmover is a combination of simple machines.

next to rivers. However, the development of the steam engine meant that machines could be built anywhere they were needed. Steam engines (and the internal-combustion gasoline and diesel engines that replaced them) relied on the crank and connecting rod to transform the up-and-down motion of a piston inside a cylinder into rotational motion that could drive a wheel.

Machines in the Modern World

For all their sophistication, modern machines today can be traced back to the simple machines developed by early civilizations. Most are little more than complex combinations of levers, gears, wheels, and axles.

The first machine guns could be used to fire a rapid stream of bullets using a hand crank. The Singer sewing machine (patented in 1850 by Isaac Singer) was powered by a foot treadle, which rotated the main drive shaft. The drive shaft then used a crank to raise and lower a needle through the cloth. The 19th-century cash register used complex arrangements of gears and borrowed the

FACTS AND FIGURES

- Cranes and hoists date from the time of ancient Greece, around 1500 BCE. Archimedes is thought to have developed a type of crane using three pulleys linked together. A similar ancient hoist known as the pentapaston had three pulleys in an upper block connected to two pulleys in a lower block, which produced a mechanical advantage of five.
- The treadmill-powered hoists used by the builders of the Roman Empire could lift loads of a few tons. The biggest crane in the world today, a Sarens SFC-250, nicknamed "Big Carl", built in Belgium in 2018, can safely lift loads of up to 5,500 tons (5,000 metric tons).

SOCIETY AND INVENTIONS

The Luddites

Even before the Industrial Revolution, groups of people were challenging the idea that machines automatically bring benefits to society. As early as 1288, French weavers were refusing to work with thread produced by mechanical spinning machines.

Perhaps the best-known opposition to machines came from a group named the Luddites in Britain during the Industrial Revolution. In 1811, this band of angry textile workers smashed up the new machines that they feared would take away their jobs. The violence spread through central England for four years. Some of the rioters were hanged.

One of the best-known comments on the dehumanization of the machine age was the film *Modern Times* (1936), in which Charlie Chaplin (1889–1977) wrestles machines on a factory production line—creating chaos in the factory.

Luddites saw factory machines as a threat and smashed them to pieces.

system of levers used in its keyboard from the typewriter. The bicycle uses the wheel and axle and an ingeniously adjustable arrangement of gears.

Although many of today's modern machines have sophisticated electronic controls, some machines are barely distinguishable from their ancient counterparts. Cranes still use hoists and pulleys that would not look out of place in ancient Rome. Archimedes' screws are still used in construction machines and on production lines. Even the automatic doors used in shopping malls owe much to the steam-powered temple doors devised by Hero of Alexandria. The difference lies in the power source. The Industrial Revolution transformed how the world makes and uses power and energy.

Yacht crews use winches and pulleys, which have been in use for centuries, to raise and lower heavy sails.

HEAT AND LIGHT

In the industrial Revolution, people discovered how to harness energy from burning fossil fuels, such as coal. Most machines are powered by energy that is released as heat from a wide range of fuels.

People are consuming ever more energy, and burning large amounts of fuel, to provide power, heat, and light.

The modern world needs a huge amount of energy to keep going. At present, most of this energy comes from fuels—substances that store energy in the way their atoms are combined together in molecules. Burning a fuel releases this energy.

Most fuels ultimately come from plants, which trap the power of the Sun's rays to build up molecules packed with energy. Wood comes from present-day plants, whereas coal and oil (also called fossil fuels) come from plants and small organisms that died millions of years ago and lay buried underground. Fuels can be solids, liquids, or gases. When scientists discovered how to release the energy from fossil fuels, it became the driving force behind world industrialization.

Solid Fuels

Wood was the most important fuel worldwide up to the second half of the 19th century, when it was overtaken by coal. In earlier times, people had also discovered that the controlled burning of wood created another fuel—charcoal. Charcoal (which is almost pure carbon) burns with a hot, smokeless flame and was long used for smelting metals from their ores, and as a constituent of gunpowder. Charcoal is less common today, but wood itself remains a useful fuel and—when obtained from managed forests—is a self-renewing one.

Coal has some of the typical disadvantages of a solid fuel. It is difficult to make it burn as fast as oil or gas. However, it is plentiful and can be used to heat water, which then turns to steam giving a reliable source of power to drive steam engines. Today, coal is mainly burned in furnaces to create the steam used in electricity generation.

COAL FIRING

Some techniques have allowed engineers to extract greater energy from coal. A method called pulverized-coal firing, introduced around 1914, is still widely used. It involves crushing the coal into tiny particles the consistency of face powder and blowing them into the furnace, where they burn like a gas flame. In 1976, another firing method, fluidized-bed combustion, was introduced. In this method, coal particles are burned within a bed of sandy material that is constantly stirred by air blasted through it—a process that usually results in less pollution than the previous coal-burning methods used.

Coal is one of the dirtiest fuels. Many impurities, such as sulfur, are mixed into it, and these have to be captured after the coal is burned to reduce pollution.

COMBUSTION

Combustion means burning. It happens when the molecules of a fuel combine with oxygen atoms in the air and release energy as heat. Natural gas, also known as methane (CH_4), combines with oxygen (O_2) molecules to make carbon dioxide (CO_2) and water (H_2O). Most fuels contain carbon, and produce carbon dioxide gas—a cause of global warming—even if no other pollutants are released. The fuels used in space rockets have to include their own oxygen source so that they will burn in outer space, where there is no oxygen to power the reaction.

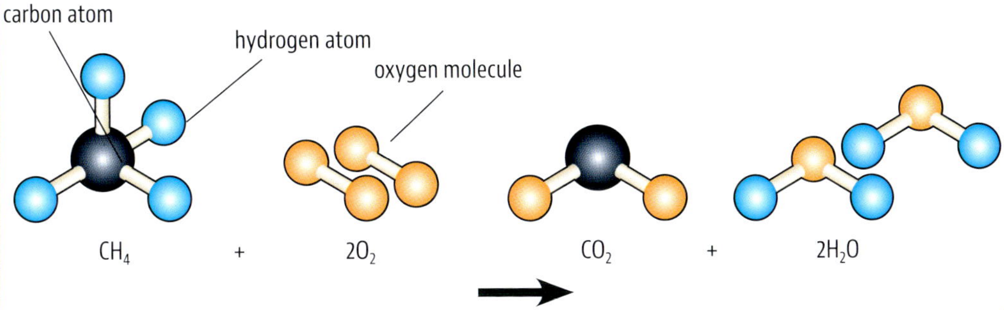

Many oil and gas reservoirs are located under the ocean floor. Oil platforms are erected at the surface to drill down through the seabed.

Oil and Liquid Fuels

Petroleum—often called crude oil or just oil—consists mainly of hydrocarbons (compounds of hydrogen and carbon). It is derived from microscopic organisms that died millions of years ago and whose buried remains were gradually converted to liquids and gases by the heat and pressure underground. Petroleum may seep to the surface naturally, evaporating to leave bitumen —a thick, tarry substance used by ancient peoples in building and medicine.

FACTS AND FIGURES

• One day fossil fuels will run out. Recent estimates suggest that at current rates of usage, oil and natural gas will run out in around 50 years. Coal is estimated to run out in about 125 years.

THE OIL INDUSTRY

Plant and animal oils have been used in oil lamps since prehistoric times. Another popular oil, kerosene, was distilled from coal and a rock called oil shale in the early 19th century. But it was only with the development of the petroleum industry in the United States in the 1850s that liquid fuels started to achieve the dominant position they hold today.

Petroleum's use as a fuel can be dated to the famous oil well drilled at Titusville, Pennsylvania, in 1859. The world's first oil pipeline followed a few years later, and the industry rapidly grew with the discovery of major oil fields in Oklahoma, Texas, and California.

Oil is stored in wooden barrels at the oil field in Titusville, Pennsylvania, in the mid-19th century. By 1872, the standard capacity of an oil barrel was set at 42 gallons (159 liters).

At first the main product from oil fields was kerosene for oil lamps and heaters, along with lubricants and tar, which was used for roads. After 1900, automobiles were manufactured in significant numbers. Gasoline, which had previously been thrown away as a worthless and dangerous product of the oil fields, became the industry's most important product. A process called fractional distillation was developed, in which dissolved gases, gasoline, diesel fuel, and heavy fuel oils (used for powering steam engines) were all separated out from the crude oil. The different components (fractions) of crude oil condense at different levels of a fractionating column and are tapped off. However, there was a limit to the proportion of gasoline that could be obtained from crude oil in this way.

Other methods were invented to convert, or "crack," heavier molecules so that they could be used as gasoline. In about 1913, thermal cracking, using heat, was introduced. Then, from 1936, more efficient methods using

FRACKING

Fracking is a method of extracting oil, especially natural gas, from Earth by drilling down and injecting a combination of water, sand, and chemicals under high pressure. This fractures ("fracking") the rock. This enables the oil and gas in the rock to be extracted and brought to the surface cheaply. Opponents of fracking argue that it contaminates ground water, leads to earthquakes, and has health implications for populations close to where fracking takes place. Supporters say it creates jobs, reduces dependence on energy imports from other countries, and lowers energy costs.

A fracking site in southwestern Pennsylvania.

Fractionating columns dominate the skyline at a refinery, separating crude oil into its constituent parts.

catalysts began to be used. Today, the refining of oil is an elaborate process that produces fuels and raw materials for the chemical industry.

Natural Gas

Like crude oil, natural gas is extracted from reservoirs under the ground. Natural gas consists mainly of flammable methane gas (CH_4). In China 2,000 years ago, natural gas was taken from the ground via bamboo pipes and burned for lighting. However, in the West, city gas (or coal gas) was the first major gas fuel to be used. City gas is a mixture of hydrogen, carbon monoxide, and methane that is released by roasting solid coal. Following experiments by Scottish engineer William Murdoch (1754–1839) and others, a company was set up to supply London with city gas in 1807, and the first gas street lights were installed that year.

Early street lights burned city gas. Lamplighters lit each lamp at dusk—and returned in the morning to turn them off.

SOCIETY AND INVENTIONS

Politics, Petroleum, and Gas

In the 21st century oil has become a powerful weapon, and countries without it often feel under threat. Germany and South Africa, both coal rich but lacking in oil, became pioneers in the technology of making oil fuels from coal. The importance of oil from the Middle East, and especially Saudi Arabia, has complicated the politics of the region.

There is also an issue with the environmental effects of drilling for oil and natural gas, especially in areas where there has been little mining activity previously, in the Arctic Ocean for example. Although fossil fuels are limited, the effects of climate change may make people to switch away from them before they run out.

Major accidents and oil spills have made the public concerned about the safety of drilling for oil, especially through deep ocean floors.

OTHER GAS PRODUCTS

Apart from natural gas, other gases or gas mixtures are also important. They are called liquefied petroleum gas (LPG) and are derived from oil refineries or by condensing from natural gas. Ethane, propane, and butane are easier to keep liquid than methane and have long been convenient fuels for remote communities. The manufactured gas acetylene (C_2H_2) burns with a very hot flame and is used for welding metals. Hydrogen is also a potentially important fuel. Unlike the other gases above, it does not contain carbon and so does not release soots and carbon dioxide, a cause of global warming, when burned. In 2025, most hydrogen was made from fossil fuels. In the future, "green hydrogen" will be made by splitting water, using electricity from solar power or other clean energy sources, which will be better for the environment.

An oxyacetylene torch uses two gas jets—one of oxygen, the other of acetylene. Combined, the gases burn hot enough to cut through metal.

A ship carries liquefied petroleum gas (LPG) in large tanks. Compressing and chilling gas fuels into a liquid makes them easier and safer to transport.

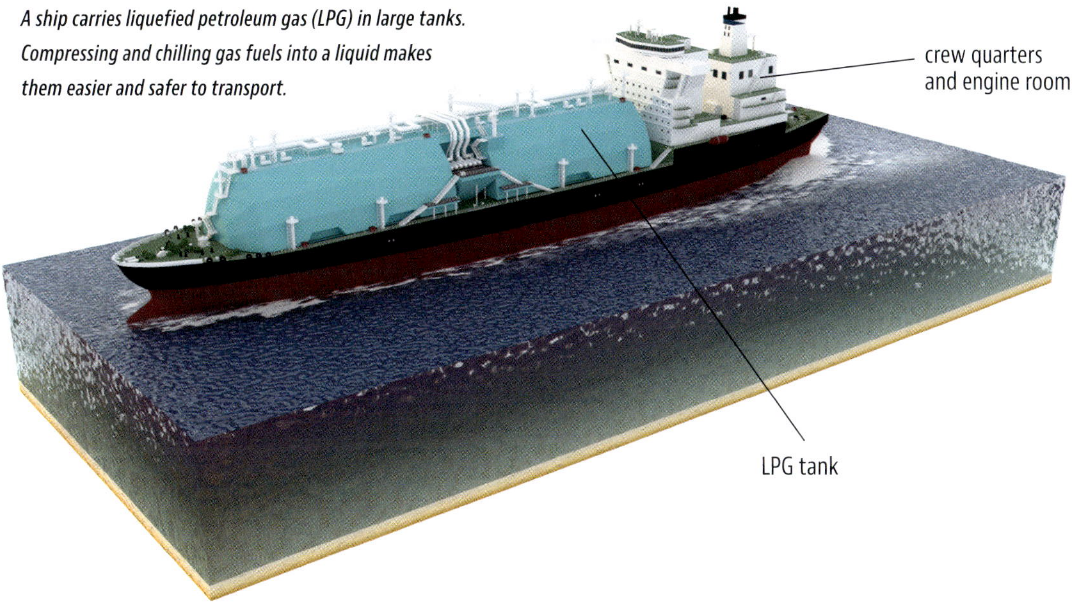

City gas was used in Europe until the 1960s, when it was replaced with natural gas. City gas is poisonous since it contains carbon monoxide. Natural gas is non-toxic, although it is still a highly dangerous substance that can easily cause explosions.

In the United States, natural gas was often found when drilling for oil, but at first no use was made of it. It has become important only in the 20th century. After World War II (1939–1945) a network of steel pipelines was built across the country for transporting gas at high pressure. Since the late 1950s, ships with refrigerated compartments have carried liquefied natural gas (LNG) across oceans. Methods of storing gas by pumping it into rock formations underground were also developed. Natural gas has increased in importance as a clean fuel suitable for use in homes.

Wood fuel was first used by the ancestors of modern humans almost 1,000,000 years ago. Even today, firewood is the main heat source for two billion people on Earth.

Heating Systems

For hundreds of thousands of years an open fire was the only form of heating. Closed fuel-burning stoves, which are more efficient, were known in ancient China and developed in the west from the 17th century. An improved version was designed by the philosopher Benjamin Franklin (1706–1790) in around 1744.

CENTRAL HEATING

An alternative to individual heat sources in a building is central heating, in which heat is generated in one place and distributed by pipes or ducts. The *hypocaust* system of the ancient Romans is an early example of this. Large Roman buildings were designed with special underfloor spaces and wall ducts through which warm air and smoke from a central furnace was circulated, so heating the rooms. Later, as steam power was developed during the Industrial Revolution, steam also began to be used for heating factories and other buildings.

Steam heating is still employed, although in modern buildings it is often replaced by warm-air ducts (especially in the United States and Canada) and circulating hot water (more common in Europe). With the invention of the thermostat around 1900 room temperatures could be controlled automatically by switching heat sources on and off, and modern systems have even more elaborate control mechanisms.

Heat pumps use underground pipes to warm a house in winter, then run in reverse to cool it in summer.

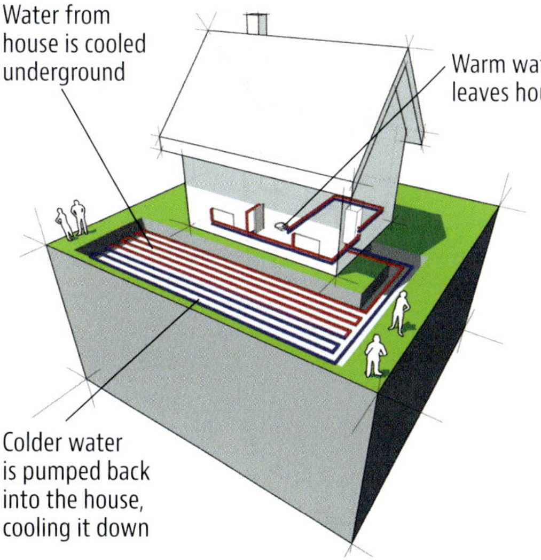

The appearance of new fuels in the 19th century led to new heat sources. Domestic kerosene-burning heaters were introduced, and toward the end of the century heaters burning city gas became common. At around the same time, the first electric heaters appeared—early examples were used in streetcars in Richmond, Virginia, in 1887. Heaters work either by radiating energy or by convection—heating the nearby air, which rises and circulates through the room—or a mixture of both. Efforts continue to make

CLIMATE CHANGE

Much of the energy from the Sun passes through Earth's atmosphere as visible light. The light is absorbed by the land and sea, warming the planet's surface. Energy is then released in the form of invisible infrared radiation, otherwise known as heat. A lot of the heat escapes into space, but some is trapped by gases in the atmosphere, which keeps Earth's air warm. This process is the Greenhouse Effect—the atmosphere works like the glass in a greenhouse, or hothouse. The main "greenhouse gases" are water vapor, carbon dioxide (CO_2), and methane (CH_4). The greenhouse effect keeps Earth's lower atmosphere warm enough for life. However, burning fossil fuels has doubled the quantity of CO_2 since the beginning of the Industrial Revolution. The extra gas traps more heat in the atmosphere, leading to a significant increase in the global temperature. Global warming will change the world's climate, creating more extreme weather, such as floods and droughts. Polar ice will melt, resulting in sea level rises that may flood coastal cities.

The world's glaciers and ice caps are melting at an ever-increasing rate, causing rising sea levels. This is caused by global warming of the planet.

better use of the energy we have. Good insulation of buildings prevents heat being wasted. Cogeneration, in which power stations produce useful heat for their local districts as well as electricity, is employed in some areas. Since around 2005, solar power has been the world's fastest growing source of electricity.

Lighting

Apart from firelight the earliest lighting devices were small, dishlike lamps. Found from prehistoric times on, these lamps burned animal and plant oils that were drawn up through an absorbent wick. Together with candles, oil lamps of various designs remained the main source of artificial light until the 19th century. In 1784, Swiss inventor François-Pierre Amis Argand (1750–1803) invented a lamp with

Thomas Edison's Ediswan company was the first manufacturer of light bulbs. The company was a partnership between Edison and Joseph Swan, who made the first light bulbs in 1878.

LASERS

"Laser" is an acronym for "light amplification by stimulated emission of radiation." Lasers are a way of focussing light, and the first functioning laser was used in 1960. They have become an essential part of the modern world, being used in everything from barcode scanners to public light displays and even deadly military weapons.

Lasers cut metal, read CDs, and help in medical procedures.

LED light bulbs are energy-efficient. They use tiny light-emitting diodes to convert electricity into light.

a wick in the shape of a ring, supplied with air on the inside as well as the outside and protected by a glass cylinder. This produced a much better light than previous lamps.

Electric lighting was developed by the 1850s. The first electric light was the arc lamp, invented in 1805 by British chemist Sir Humphry Davy (1778–1829). Davy made an electric spark jump between two carbon rods, creating an intense light. Inventors began to explore the idea of making thin filaments (strands) glow by using an electric current.

A near vacuum would be needed around the filaments to keep them from burning. The result, in 1878, was the light bulb, or incandescent lamp, developed by Thomas Edison (1847–1931) and others. Early filaments were made from carbon, but were later replaced by tungsten metal, which glows hotter and brighter without melting. Through the 20th and 21st centuries, incandescent lamps have gradually been replaced by more energy-efficient compact fluorescent lamps and LEDs (light-emitting diodes).

GAS MANTLE

After city gas came into use around 1800, many designs for gas burners were marketed. Attempts were also made to place various materials in the path of the gas flame so that they would glow and make a much brighter light. The result was the gas mantle invented in 1885 by Carl Auer von Welsbach (1858–1929). A mantle is a cloth net coated in chemicals that glow when heated.

Camping lanterns are fitted with a cloth mantle that converts the flame into a brilliant light.

FLUORESCENT STRIP LAMP

A typical fluorescent light is a tube filled with mercury vapor at very low density with electrodes (electrical terminals) at each end. When the light is switched on, electrons are released from the negative electrode (the cathode) and move toward the positive one (the anode). On the way they bump into mercury atoms, making them give off ultraviolet (UV) light. This UV light is absorbed by the special fluorescent coating (containing phosphor crystals) on the glass of the tube, and the energy is emitted as visible light. Fluorescent lights use energy more efficiently than incandescent bulbs and are common in large buildings.

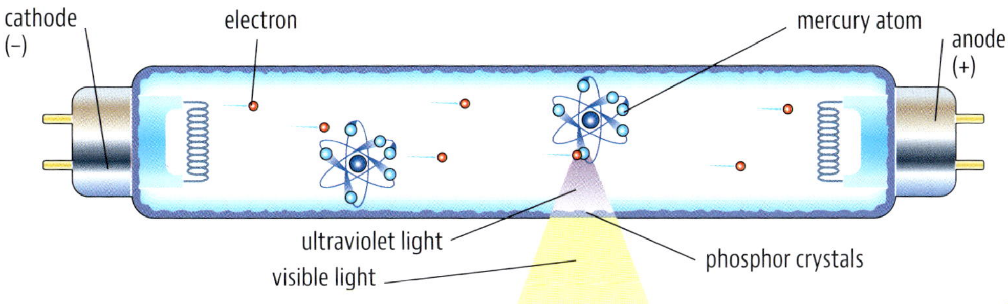

THE DRIVE FOR POWER

This car engine is a combination of pipes, chambers, and levers that work to extract energy from gasoline to turn the car's wheels.

Engines drive the modern age. From airplanes and locomotives to rockets and cars, almost all are powered by engines that burn fuel. However, greener sources of power are now being used to drive cleaner machines.

The word engine was once used to describe almost any kind of machine, including a water mill. Strictly speaking, an engine that burns fuel to produce heat and converts that energy

GAS PRESSURE

Gases, such as steam, exert pressure as the gas molecules move around and collide with surfaces. Each time they collide with the walls of a container, they push against them. Pressure is measured as the amount of pushes, or force, in an area. Adding more gas to the container increases the collisions and ups the pressure. Similarly, when gas is heated, the molecules move more quickly, collide more often, and exert higher pressure.

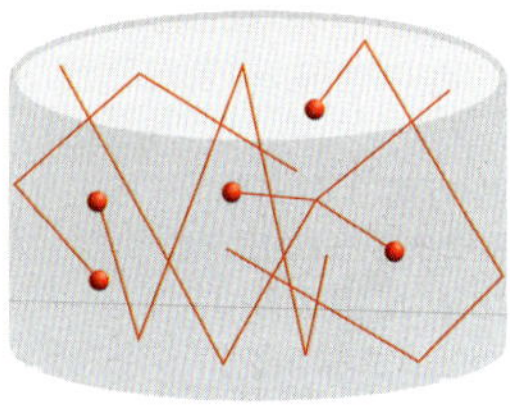

Gas hits container walls, exerting pressure

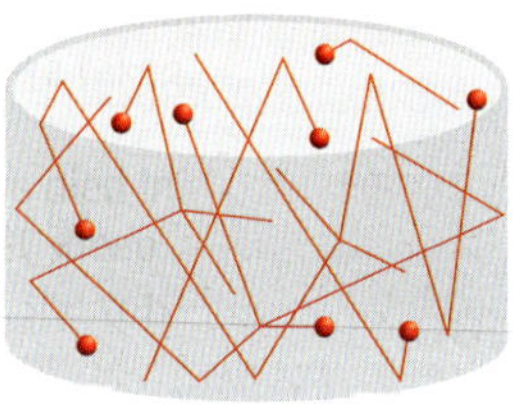

More gas in the same space exerts higher pressure

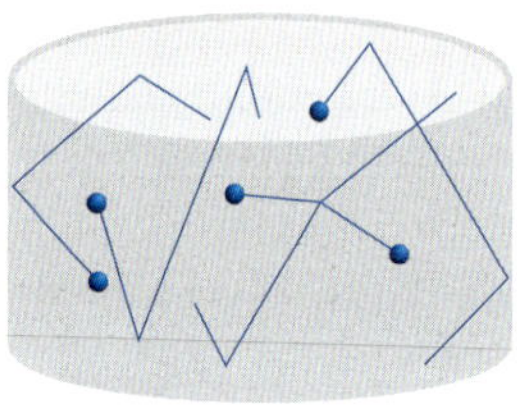

Cooler gas moves more slowly and exerts lower pressure

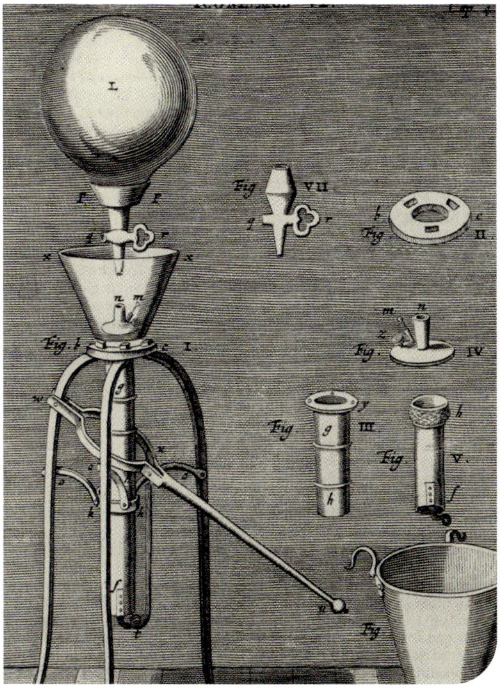

In the 17th century, Robert Boyle's research into the behavior of gases relied on the development of advanced pumps, as in this historical diagram, for compressing gases to high pressures.

into useful motion is called a heat engine. The first steam engines heated water to produce steam; they were followed by engines that burned gasoline and diesel. Some engines use hydrogen or other "clean" fuels. Others have been replaced by electric motors.

The Long Road to Steam

It was not until the 17th century that serious efforts were made to harness the power of steam. In 1660, English physicist and chemist Robert Boyle (1627–1691) came up with a series of laws to describe the behavior of gases. These gas laws included the important observation that if a gas is heated in a closed container, its pressure increases. This principle has been used in every engine made since that time. In 1679, French mathematician and physicist Denis Papin (1647– 1712), who had worked in England for Boyle, invented a small steam-powered device that he hoped would pump water from deep mines. However, it failed because Papin combined the boiler and the cylinder in a single device. It was three decades later, in 1712, that the first effective steam engine, built by British inventor Thomas Newcomen (1663–1729), was successfully used to pump water from a mine. Over 1,000 Newcomen engines were built.

THE AEOLIPILE

Although the first steam engines were not developed until the 18th century, experiments with steam as a form of power date back as far as Hero of Alexandria, a Greek inventor who lived in the first century CE. His *aeolipile* was a sphere with two exhaust nozzles mounted on opposite sides. When water was boiled, it filled the sphere with steam, which emerged from the nozzles, causing the sphere to spin around until all the steam had been exhausted.

As far as is known, the motion of Hero's aeolipile *was never harnessed to power machines. It was merely displayed as a novelty.*

Today, Newcomen's steam engine seems slow and inefficient. It was a single cylinder, which had to be alternately heated and cooled. However, the problem was solved by another British inventor, James Watt (1736–1819), who worked as an instrument maker at the University of Glasgow in Scotland. Watt added a separate cylinder (called the condenser) to his engine that could be constantly cooled. This was much more efficient than Newcomen's design, reducing the cost of running steam engines by around 75 percent.

Although Watt's first engines could produce only reciprocating (up-and-down) motion, his partner, British engineer and industrialist Matthew Boulton (1728–1809), persuaded him to design engines that could produce rotary (turning around an axis) motion. Watt's later designs were beam engines, in which the piston in the cylinder rocked a large beam up and down, and the beam turned a wheel using a system of gears. He invented the double-acting rotative steam engine in 1782. In this model, steam pressure was sent alternately to each end of the cylinder. For safety reasons Watt's engines used steam at low pressure, but this meant they had to be very large, and so restricted what they could be used for.

James Watt did not invent steam engines but his innovations made them much more practical for industrial use. The unit of power, the Watt (W), was named for him.

NEWCOMEN'S ENGINE

Thomas Newcomen's steam engine consisted of a large copper boiler in which water was heated by a coal fire to produce steam. His innovation was to introduce a separate piston (a movable plunger) and cylinder above the boiler that converted the expansion of steam into useful up-and-down (reciprocating) motion. Steam from the boiler was allowed to expand into the cylinder, pushing the piston upward and driving the pump. The cylinder was then cooled by water, creating a partial vacuum, which pulled the piston back down and drove the pump the other way.

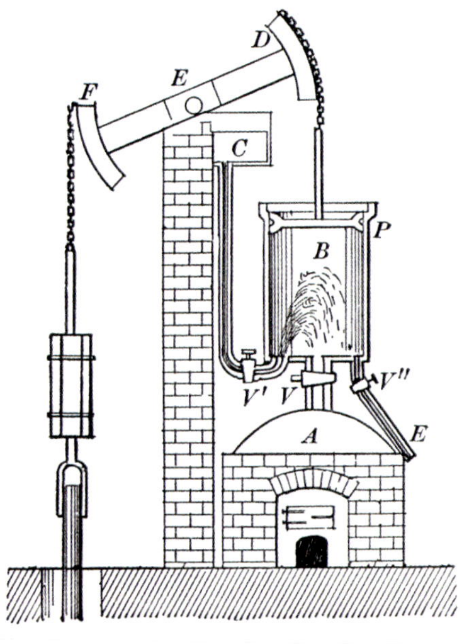

Newcomen's "Atmospheric Engine" was said to have the power of 20 horses. Horsepower was used to compare the strength of later engines and is still the unit used today.

STEAM ENGINE

Steam engines became more efficient using Watt's two pistons. Steam from the engine's boiler enters the engine's cylinders through a system of valves. The steam puts pressure on the main piston, pushing it toward the end of the cylinder. This pulls the connecting rod toward the engine and the crank (the link to the flywheel) converts the engine's reciprocating motion into the rotary motion of the flywheel. As the flywheel rotates, the cam-shaped eccentric strap also rotates, and the wide part of the eccentric strap pushes the eccentric rod toward the engine. The eccentric rod is linked to the smaller piston, and the piston is pushed to the end of the cylinder. That directs the steam into the opposite end of the main cylinder, forcing the main piston back and reversing the direction of the connecting rod. Thanks to the crank, the flywheel keeps turning clockwise. The eccentric strap rotates, pulling the smaller piston back to its starting position, restarting the cycle.

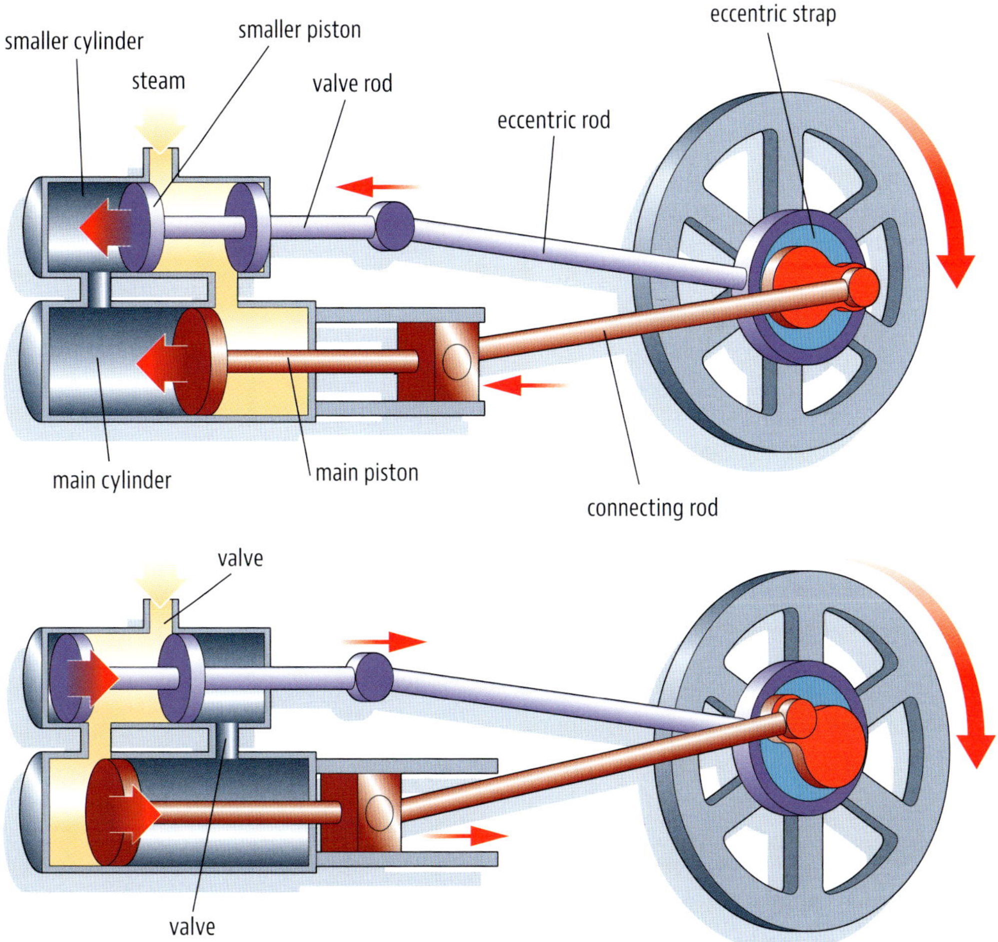

The American engineer Oliver Evans (1755–1819) developed a smaller, high-pressure steam engine than Watt's designs, and produced around 100 different engines for use in factories. In 1801, another pioneer of the high-pressure engine, British engineer Richard Trevithick (1771–1833), produced a steam-powered vehicle for carrying passengers by road—this was the ancestor of the automobile.

In a locomotive, hot smoke from the firebox, or furnace, passes through pipes (yellow in this illustration) running through the boiler. The pipes heat the water to form steam, and the smoke escapes through the smokestack, or funnel.

Internal Combustion

Despite their popularity, the new steam engines had many disadvantages. They were dirty, noisy, heavy, and large, and they needed separate supplies of both water and fuel. Largely because of the separate boiler needed to heat the water, they were also very inefficient. This type of external-combustion

THE STIRLING ENGINE

The worst thing about early steam engines was that their boilers had a tendency to explode at very high pressures. In 1816, in an attempt to solve this, British inventor Robert Stirling (1790–1878), devised a form of external-combustion engine that could operate without a boiler. The Stirling engine used two pistons linked together inside a single cylinder, which was heated externally by a furnace. One piston was used to drive a wheel, while the other piston was used to expel hot air. This engine can run on any kind of fuel, from wood to nuclear power. Stirling engines create less pollution than ordinary engines and are also highly efficient. They can also run on gasoline, but they are usually too big to fit into automobiles.

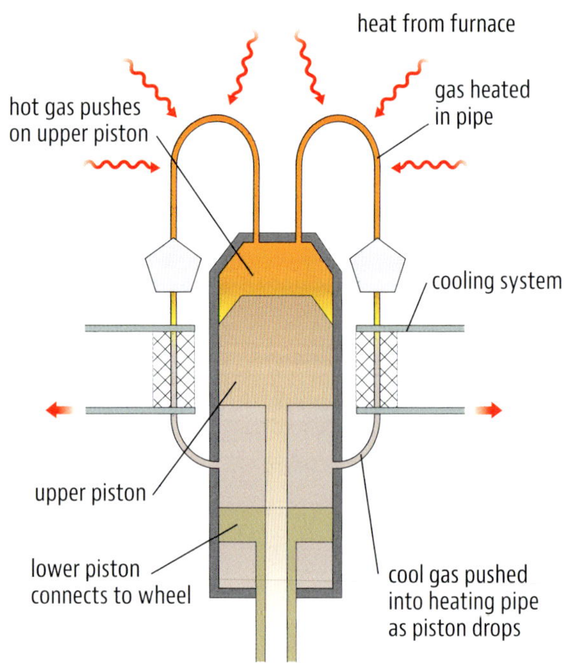

SOCIETY AND INVENTIONS

Traveling By Steam

By 1810, the power of steam had made it possible for people to travel in a variety of new ways. Richard Trevithick had produced the steam-powered carriage in 1801. By 1804 he was using similar technology to power the world's first steam locomotive. Steam engines soon found their way into ships. The first real steamboat was constructed in 1801 by British mechanical engineer William Symington (1764–1831). By 1807, the American engineer Robert Fulton (1765–1815) had started a regular steamboat service along the Hudson River between New York and Albany. His steamboat *Clermont* traveled at an average speed of 5 mph (8 km/h). Steam power made it possible for people to travel by road, rail, or sea, but there was always the constant danger of a boiler explosion.

The SS Great Eastern, *an iron steamship launched in 1858. It dwarfed other ships being constructed at the time.* Great Eastern *remained the world's largest ship for 41 years.*

engine (in which the fuel is burned outside the main cylinder) gradually gave way to a much more efficient internal-combustion design, in which the fuel is burned in the cylinder itself.

Engines that used internal combustion emerged toward the end of the 19th century after research by French engineers Nicolas Sadi Carnot (1796–1832) and Alphonse Beau de Rochas (1815–1893). In 1824, Carnot published a theory of engine efficiency using Robert Boyle's gas laws, which he called *Reflections on the Motive Power of Heat*. However, so much practical experimentation with steam was already taking place that Carnot's work made little impression. In 1859, French engineer Jean Joseph Lenoir (1822–1900) built an engine that worked by burning a mixture of city gas and air. In 1862, De Rochas demonstrated that a practical engine would need to pass through a four-stage cycle of operation.

By 1867, German engineer Nikolaus August Otto (1832–1891) had produced the first working internal- combustion engine that used the four-stage (or four-stroke) cycle. It is not clear whether Otto's engine was based on the earlier work of De Rochas in France, but the invention of the engine that uses the four-stroke cycle is credited to Otto. Almost every automobile engine ever built has used this idea.

Using Gasoline

The Otto engine was a considerable advance on the steam engine, but it still had its problems. Chief among these was the fuel it used; gas was difficult to supply and very costly. In the 1870s and 1880s engines were produced that used other fuels, including kerosene. Another German engineer, Gottlieb Daimler (1834–1900), developed the familiar gasoline engine in 1884. His innovation was to introduce a separate chamber, known as a carburetor, in which gasoline was mixed with air to form an explosive mixture. Another major development occurred in 1892 when a German mechanical engineer called Rudolf Diesel (1858–1913) patented the use of heavy oil (now named for Diesel) in an engine based on the Otto cycle. For all the refinements made in the last century, today's engines are

From the earliest days, car manufacturers have shown off their expertise by building racing cars fitted with powerful engines.

still visibly related to those early, four-stroke, piston engines produced by Otto, Diesel, and Daimler. While piston engines have their advantages, they are not without their drawbacks. Most obviously, the power they can produce is limited: cylinders can be made only so big; pistons can move only so quickly; and there is a limit to how many pistons and cylinders an engine can have without becoming too heavy. The engine that offered a solution was the gas turbine. Unlike a steam turbine, which is turned by steam from a separate boiler, a gas turbine is turned by hot gases that escape when fuel is burned inside the engine chamber. A steam turbine is an external-combustion engine, but a gas turbine is an internal-combustion engine.

British inventor John Barber (1734–1801) patented a theoretical device featuring the main components of a gas turbine in 1791, but the first successful engines were not built until around 1903 in France.

DIESEL ENGINES

The diesel engine does not use spark plugs; instead, it compresses the fuel to much greater pressures so that it ignites by itself. When invented in the 1890s, the diesel engine was the most efficient form of engine yet made, but to achieve the higher pressures it had to be both heavier and sturdier. This explains why, historically, diesel engines have always been favored for trucks, ships, and larger forms of transportation, and why gasoline engines have typically been used in automobiles.

FOUR-STROKE INTERNAL-COMBUSTION ENGINE

In an internal-combustion engine, the piston is driven by fuel burning in the cylinder itself, forcing a piston to move. The piston moves up and down inside the cylinder as the engine repeats a cycle of four separate stages, or strokes.

1. Induction The piston moves down, allowing a mixture of fuel and air to enter the cylinder through the inlet valve.

2. Compression The piston moves up, compressing the fuel and air mixture, so it is ready to burn.

3. Ignition The mixture is ignited by an electrical spark plug, causing an explosion in the cylinder. The burning mixture heats up and expands, pushing the piston down.

4. Exhaust The piston moves up and pushes the burned gases out through an exhaust valve.

In most car engines four or more pistons rotate a crankshaft **5**. The "big end" of each connecting rod is attached to the crankshaft, which converts the reciprocating motion of the pistons into rotary motion.

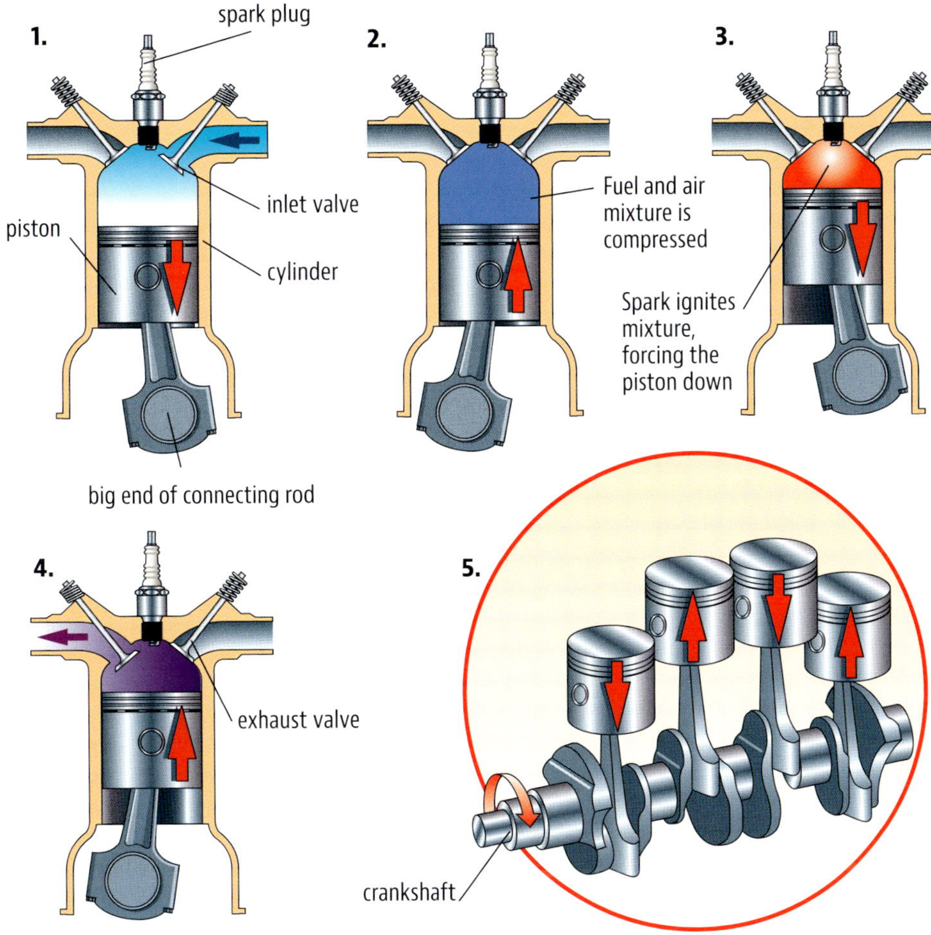

FACTS AND FIGURES

- American engineers Wilbur (1867–1912) and Orville (1871–1948) Wright, brothers who produced the first practical airplane. They built their own four-cylinder internal-combustion engine in 1903. Cast from aluminum, it weighed just 152 lb (69 kg) and produced around 12 horsepower at 1,000 revolutions per minute. It drove the Wright's pioneering airplane with a modified bicycle chain.
- Modern airliners are powered by engines that weigh around 13,000 lb (6,000 kg) and produce a thrust of 70,000 lb (310,000 N).
- The biggest jet engine is the General Electric GE9X. It is 11 feet (3.4 m) in diameter and produces 134,300 lb (597,396 N) of thrust.

Propeller-powered aircraft are driven by radial piston engines. The pistons, powered by internal combustion, are arranged in a ring around the central crankshaft.

SIR CHARLES ALGERNON PARSONS

In 1884, British engineer Charles Parsons (1854–1931) pioneered high-speed steam turbines, in which a jet of steam from a boiler turns a continuously rotating paddle wheel instead of driving a piston. Parsons went on to develop the first practical steam turbine and pioneer its use in power plants and ships. His steam-turbine yacht *Turbinia* (right), set sail in 1887. By 1910, warships were being powered using his technology.

In the United States gas turbines were developed later than in Europe and initially for industrial use. A young American engineer named Sanford H. Moss worked on their design at Cornell University from 1902 and later produced military devices for General Electric. In 1937, a Swiss engineering firm called Brown Boveri developed the first industrial gas turbine for the Sun Oil Company in Pennsylvania. Union Pacific Railroad developed gas turbines for its locomotives in 1941 and within a decade had converted its fleet to the new engines. However, gas turbines are most suited to producing large amounts of power over a sustained period of time. They therefore proved very expensive in terms of fuel consumption, and the Union Pacific Railroad had abandoned them by 1969.

GAS TURBINES

A turbine is a wheel with fins or blades similar to a propeller or an electric fan. The blades are turned when a liquid or gas flows past them. A water wheel is an example of a simple turbine.

A gas-turbine engine has three key components: a compressor, a chamber where fuel is burned (the combustion chamber), and a rotating turbine. Air is sucked through an inlet and compressed by the compressor, before being mixed with fuel in the combustion chamber, where the mixture is burned at a very high temperature. Typically, gas leaves the combustion chamber at up to 3,000°F (1,700°C) and speeds of up to 1,800 mph (2,900 km/h). The hot gas rushes past the turbine, making it turn at high speed. The turbine can be used to drive a propeller (in a turboprop engine), a locomotive's transmission, or a generator in a power plant to produce electricity. In a jet engine the turbine simply turns the compressor, and instead of driving a propeller, the hot gas is thrown back through the engine to propel the aircraft forward. Lift is provided by the wings. The main advantage of a gas turbine over a piston engine is that combustion takes place continuously: much more fuel is burned, at a much higher temperature, in a much shorter time. It is this that makes a gas turbine powerful enough to drive a jumbo jet or a railroad locomotive.

The blades of a turbine are shaped to harness the forward movement of a fast-flowing fluid (in this case steam). They also change its direction into a rotation that can be used to drive a machine.

Just as steam turbines replaced steam engines in boats and ships, so gas turbines eventually replaced steam turbines. In 1947, the Motor Gunboat 2009 became the first warship to have a gas turbine. Within a decade merchant ships were also being fitted with gas-turbine engines. Today, gas turbines are still used extensively in warships and larger vessels.

Two thick pipes bring high-pressure steam to a large turbine used to generate electricity in a power plant.

SCIENCE WORDS

Combustion: The chemical reaction that happens when a substance burns in air.
Ion: An atom that has lost or gained electrons and taken on an electric charge.
Turbine: A series of fans that spins when gases—or liquids—flow through it.

ELECTRIC CARS

The polluting effects of traditional automobile engines and limited stocks of fossil fuels mean that electricity will probably be the major power source for automobiles from the late 2020s. Many countries have plans to phase out gasoline and diesel engines and replace them with electric motors powered by renewable energy from the Sun or from the wind. Hybrid vehicles switch between a gasoline engine and an electric motor to suit the driving conditions. Electric cars have advantages in the power they can generate, but the distance they can travel is limited by battery life. A small fraction of cars are powered by electricity alone.

An electric car, built by Thomas Parker, was photographed in 1895. However, it was not commercially viable at the time. Electric cars are now the future of motoring.

FUEL CELLS: ENGINES OF THE FUTURE?

A fuel cell produces electricity from a chemical reaction between hydrogen and oxygen that makes water. Fuel cells work a little like batteries, but they do not go dead as long as the chemical fuels continue to be pumped into them. The fuel cell has two electrodes: the anode and cathode. These are immersed in an electrolyte, a liquid that contains charged particles. At the anode, the supply of hydrogen atoms turns into a stream of charged hydrogen ions, which travels through the electrolyte to the cathode. There, hydrogen ions react with a supply of oxygen to form water. The constant movement of the hydrogen ions creates a current of electricity, which is used to drive an electric motor.

Fuel cells are not engines. They are more like mini power plants, generating electricity to power vehicles or any other machine. Fuel cells run very efficiently, but making pure hydrogen to fuel them requires a lot of energy. The main source of hydrogen is water, gasoline, or natural gas, and the purification processes creates pollution, even if the fuel cell itself does not. Until these fuel problems are solved, fuel cells will not be as clean and efficient as they seem.

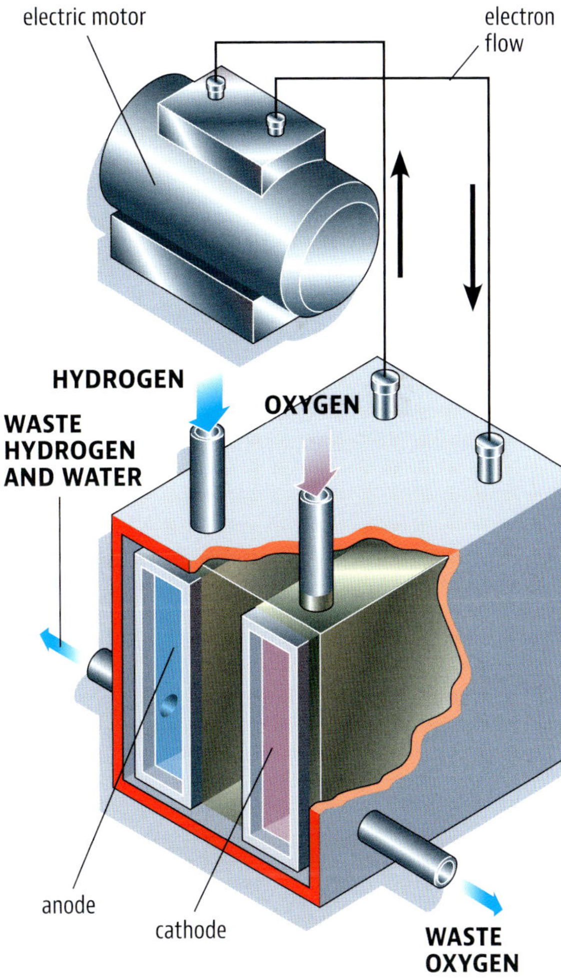

HARNESSING ELECTRICITY

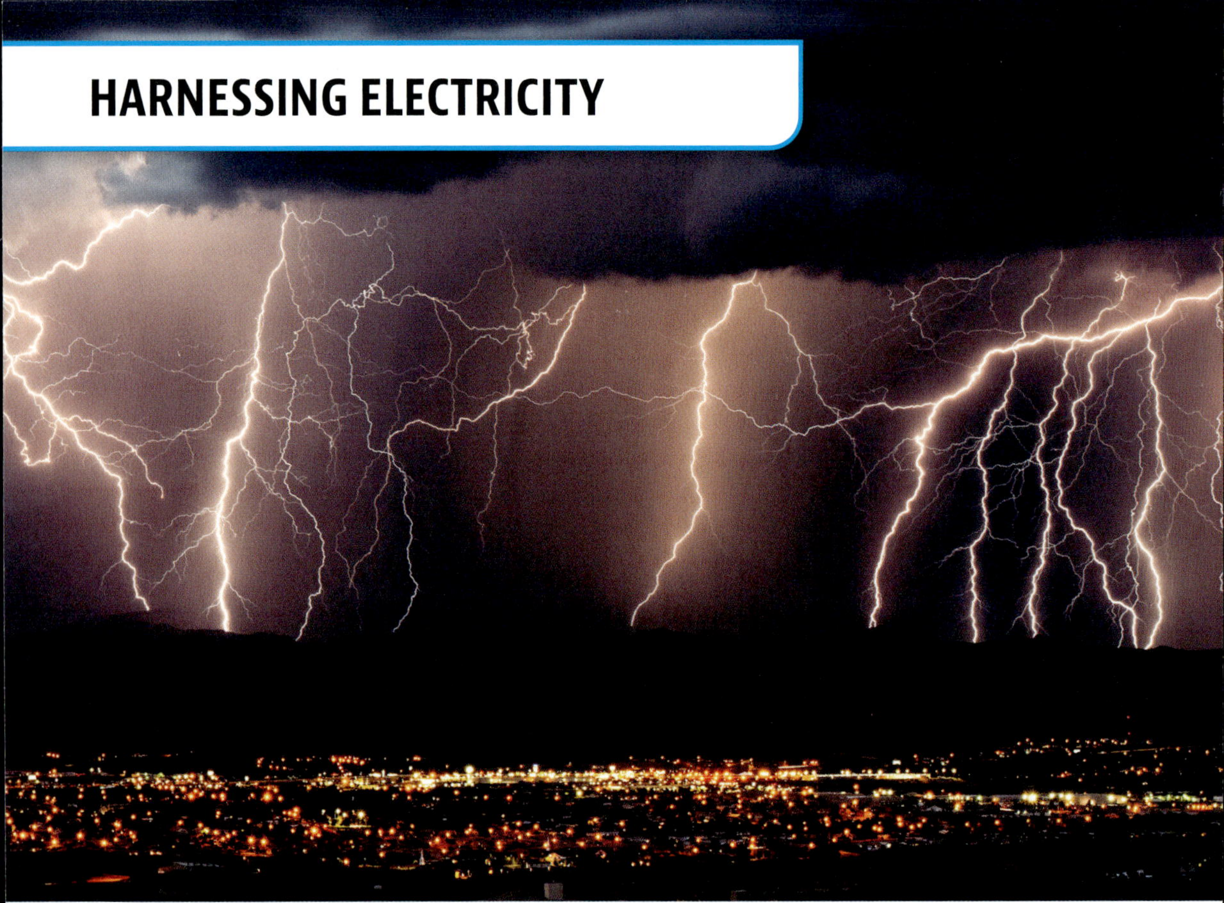

Lightning is a natural source of electricity. Each bolt contains enough energy to power a modern LED lamp for about three years, and 100 bolts strike Earth every second!

If the people of ancient civilizations could travel forward in time, there is one modern invention in particular that would seem like magic to them—electricity.

Of all the inventions associated with power and energy, electricity is perhaps the one that has changed society the most. Machines made life easier, and engines allowed them to be powered from a variety of different fuels wherever they were needed. But electricity has allowed machines and their sources of power to be separated by vast distances. A power station can be thousands of miles from the kettle that it causes to boil.

Deposits of amber. The Greek word for amber, elektron, *gives us the words electricity and electron (the particle that carries electricity).*

STATIC AND CURRENT ELECTRICITY

Electrons are tiny, negatively charged particles that, along with protons and neutrons, make up atoms. They can also exist independently and are responsible for most of the phenomena of electricity.

Static electricity is caused by a buildup of electric charge. If you walk across a carpet, friction between your shoes and the fibers causes electrons to move from your body to the carpet. This makes you positively charged. The more you walk, the bigger the charge you build up. Eventually, if you touch a metal object the charge flows rapidly back to the ground and can give you a mild electric shock. Lightning happens in much the same way when clouds pass through the air.

The Van de Graaff generator, invented by the American physicist Robert Van de Graaff (1901–1967), is a machine that can produce over a million volts of static electricity for use in particle physics experiments. Friction caused by a comb at the base of the machine loads an upward-moving conveyer belt with positive charges that collect in the hollow, metal sphere at the top.

If static electricity is a buildup of electrons in one place, current electricity is caused when the electrons flow around a closed path called a circuit. A lightning bolt is a crude type of circuit between a cloud and Earth. The bulb, battery, and switch in a flashlight form a simple circuit. When the switch is "on," the circuit is completed, power is supplied, and the bulb lights up.

VAN DE GRAAFF GENERATOR

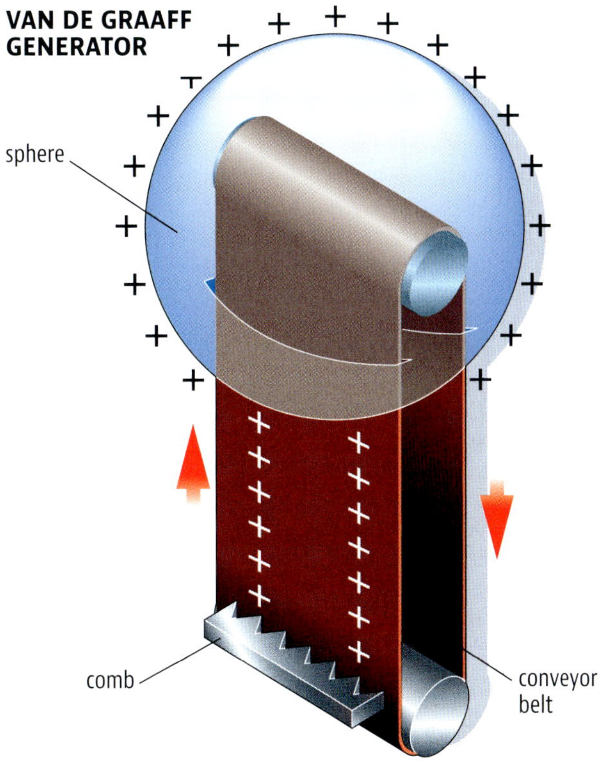

FLASHLIGHT CIRCUIT

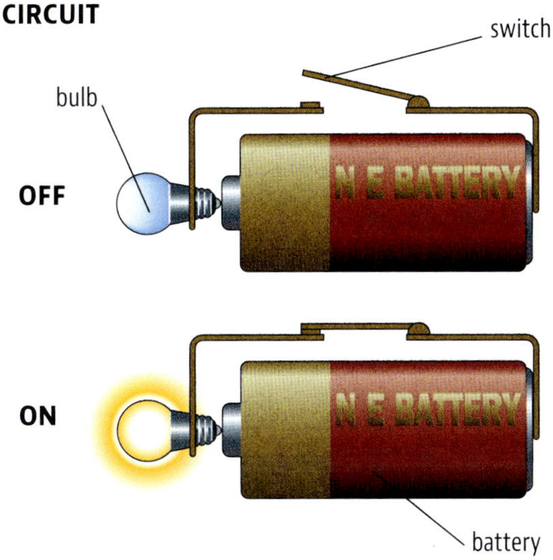

PROPERTIES OF ELECTRICITY

When an object is lifted from the ground, it gains energy in the form of "gravitational potential." If the object is then released, this energy is given out as it falls from a high potential to a lower one (hitting the ground). Similarly, the function of a battery is to raise the electrical potential at one of its terminals relative to the other.

This causes electrons to flow from high to low potential around the circuit. The difference in potential between the terminals is called the "potential difference" and is expressed in volts. A typical modern battery has an output of c. 1.5 volts. Overhead power cables carry 400,000 volts.

There are two forms of electric current, direct current (DC) and alternating current (AC). In a DC circuit, current flows in one direction; in an AC circuit the current changes direction, flowing back and forth more than 100 times each second. The two forms have different uses: DC is used by sensitive electronic equipment and in applications such as electroplating; AC is used to carry electricity from power stations because it can be generated and transmitted more efficiently. Batteries produce DC, while the wall sockets in your home provide AC.

Sparks fly between electric wires when the voltage is high enough for the current to pass through air. Air is a much poorer conductor than metals.

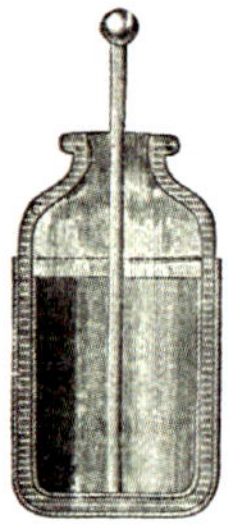

Pieter van Musschenbroek (1692–1761) invented the Leyden jar, which was the first device for storing electric charge. The jar had an inner and outer coating of metal that held the electric charge.

Forms of Electricity

We are most familiar with electricity in the form of electric currents, which pump around circuits driving televisions, PCs, and stoves. However, the scientific pioneers only knew about static electricity, the form that causes lightning and the sudden shocks built up from walking on carpets. The grandfather of static electricity was probably Greek philosopher Thales of Miletus (c. 624–546 BCE), who discovered in the sixth century BCE that if he rubbed a type of fossilized tree resin called amber, it would attract objects such as straw and feathers.

It was not until the 17th century that static electricity, or electrostatics, came to be understood. Around 1600, the English physicist and doctor William Gilbert (1544–1603) suggested electricity was caused by a fluid called humor, which could be removed from a piece of amber when it was rubbed. German physicist and engineer Otto von Guericke (1602–1686) built a primitive electric generator consisting of a globe of sulfur that built up a large charge when it was rubbed.

Currents and their Effects

In 1733, French chemist Charles du Fay (1698–1739) showed that electric charges came in two varieties: "vitreous," which we would call a positive charge today; and "resinous," or a negative charge. Although the ancient Greeks knew about static electricity (or at least the phenomena it caused), it was not until 2,000 years later that current electricity was discovered. In 1780 an Italian professor of anatomy, Luigi Galvani (1737–1798), was experimenting with the effects of electricity on dead fish and frogs. He found that an electric charge would make the muscles of a frog's leg contract, causing the leg to twitch. He concluded that this was caused by something inside the muscle. Alessandro Volta (1745–1827), who was professor of natural philosophy at a nearby university, believed

BENJAMIN FRANKLIN

One of the most important figures in the early days of electricity was the American statesman, publisher, and philosopher Benjamin Franklin (1706–1790). He was so captivated by the magic of electricity that he sold his printing business in order to devote his life to its study.

Today, Franklin's best-known invention is probably the lightning rod, or conductor. He demonstrated its power by flying a kite in a thunderstorm, catching the bolts of lightning and holding them in a charge-storing sandwich of glass and metal foil called a Leyden jar. He was lucky not to have been killed.

Unlike Charles du Fay, Franklin suggested electric charge was caused by the presence or absence of a single fluid. Franklin created the terms positive (having too much fluid) and negative (having too little). Today, negative charge is taken to mean having too many negatively charged electrons, and positive charge means having too few.

Benjamin Franklin is said to have run a wire up a kite string and attached it to a metal key. This carried the electric charge from the lightning to a Leyden jar, a primitive type of capacitor or storage unit for electricity.

Alessandro Volta invented the electric pile, or battery. The force pushing electric currents along is known as voltage, and measured in Volts (V), in his honor.

that the effect was caused by the flow of electricity around a circuit. Between them they had discovered electric current. Volta found he could produce electric currents with a device made from alternate layers of silver, leather, or pasteboard (dried wood pulp) that had been soaked in salt water, and zinc. He had produced the world's first battery, which was named a Voltaic pile for its inventor.

In a battery, a chemical reaction produces an electric current, but the reverse can also happen: an electric current can be used to produce a chemical reaction, in a process called electrolysis. Several British scientists, including the chemist Sir Humphry Davy (1778–1829), pioneered this science, later called electrochemistry, to isolate new elements. It also gave rise to the electroplating technique, where one metal can be coated with a layer of a different metal.

During the first half of the 19th century scientists around the world devoted

THE BATTERY

A battery uses a chemical reaction to generate electricity. The materials in the positive and negative electrodes would react if mixed by exchanging ions (atoms that have become charged). However, the battery keeps the electrodes separate using an electrolyte paste. The paste does not stop the reaction, but forces the ions to travel through it between the reactants, thus creating an electric current that can be used for power.

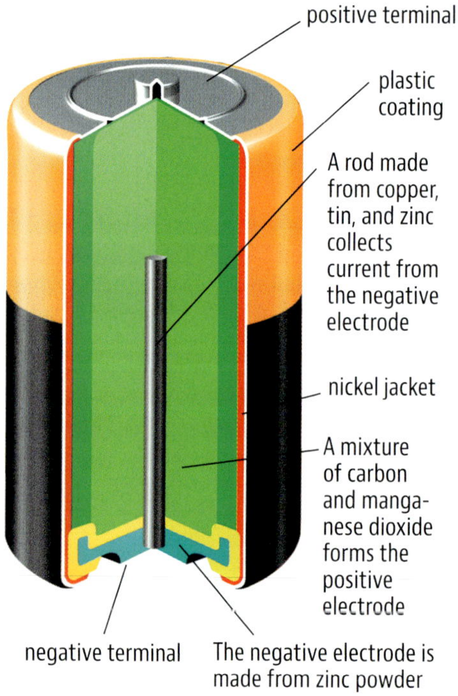

MICHAEL FARADAY

British electrical pioneer Michael Faraday (1791–1867) had almost no formal education. As a teenager he became an assistant to Humphry Davy, and when Davy went on a long European tour, Faraday went with him. It gave him the opportunity to meet most of the respected physicists and chemists who were trying to unravel the mysteries of electricity. Although his early research concentrated on the study of static electricity, Faraday carried out a famous experiment in 1831, which secured his place in the history of science. He rotated a copper disk between the poles of a large magnet and found that it produced an electric current. Faraday had invented the generator, a device that eventually made it possible to produce electric power on a large scale.

Michael Faraday's discoveries were the subject of the first Christmas Lecture at the Royal Institution in 1856. Top scientists have given a Christmas Lecture at the Institution almost every year since then.

SUPERCONDUCTORS

Insulators are materials, such as wood and rubber, that do not conduct electricity. Materials that allow electricity to pass through easily, such as silver and copper, are conductors; they are said to have a low resistance to electricity. However, even the best conductors offer some resistance.

When a conductor of electricity is cooled to a very low temperature, near the lowest theoretically possible temperature of −460°F (−273°C) or absolute zero, its resistance seems to vanish magically. This effect, called superconductivity, is one of the most exciting areas of research into electricity. It was discovered in 1911 by Dutch physicist Heike Kammerlingh Onnes (1853–1926). So-called high-temperature superconductors were developed in the mid-1980s, which show the same effects at temperatures of around −240°F (−150°C).

Superconductivity might be used in ultrafast computers, and in the development of efficient methods of transmitting electricity over long distances with little or no loss of power. It is currently being used in the electromagnets that make high-speed, magnetic levitation (maglev) trains. These trains "float" in the air and race along at up to 375 mph (600 km/h).

A superconducting magnet levitates on the repulsive force of a larger magnet.

TRANSFORMERS

A transformer is a device for changing the voltage of an electric current. It has two wires wrapped around an iron core. One coil has more turns than the other. A current in the red coil induces another in the blue one with twice the voltage. In the reverse direction, the voltage is halved. Transformers are used in the plugs of low-voltage appliances.

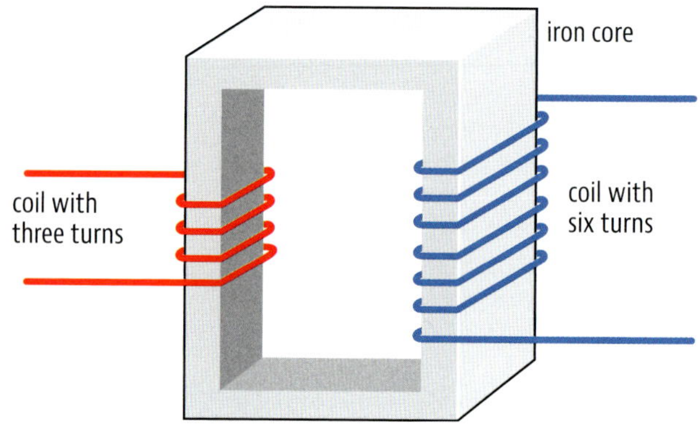

themselves to conquering the mysteries of electricity using the new batteries to power their experiments. German physicist Georg Simon Ohm (1789–1854) showed that when different materials conduct electricity, the amount of resistance they offer to the passage of a current depends on their size and the material from which they are made. The British physicist James Joule (1818–1889) demonstrated how a particular quantity of electric current could be used to produce an equivalent amount of heat—a property now used in all kinds of electric appliances from stoves to irons.

A multimeter is used by electricians to measure three aspects of electricity: voltage, resistance, and current.

Electricity and Magnetism

Perhaps most important was the research that showed how electricity and magnetism were two different sides of the same thing. In 1820, Hans Christian Ørsted (1777–1851), a professor of physics in Denmark, showed that an electric current flowing through a wire would deflect a nearby compass needle. This effect demonstrated that an electric current could produce magnetism. French physicist André-Marie Ampère (1775–1836) subsequently worked out the mathematical theory behind Ørsted's findings. Because this was such an important theory, electric current is now measured in units called amperes, or amps.

If electricity could produce magnetism, could magnetism produce electricity? This principle—known as electromagnetic induction—was confirmed separately by British physicist and chemist Michael Faraday and the American physicist Joseph Henry (1797–1878) between 1820 and 1831. These findings were combined in an elegant theory

MICROPHONES AND LOUDSPEAKERS

Microphones and loudspeakers are examples of transducers. These are devices that convert energy from one form to another. Sound waves carry energy through the air as a regular pattern of disturbancein the air pressure.

A microphone can convert this sound energy into electrical energy. The sound waves push against a piece of plastic called a membrane or diaphragm, which in turn exerts pressure on a piezoelectric crystal. When a force is applied to crystals of this type, they produce an electric charge. The charge flows through a circuit inside the microphone as an electric current. The strength of the current depends on the amount of energy in the original sound waves.

In a loudspeaker the opposite happens. An electric current generated by a radio, stereo, or TV is used to power a coil of wire, generating a magnetic field. The coil is placed between the poles of a large, permanent magnet, where it vibrates as the current flows through it. A large paper cone attached to the coil also vibrates, producing the variations in air pressure that are then heard as sounds being reproduced.

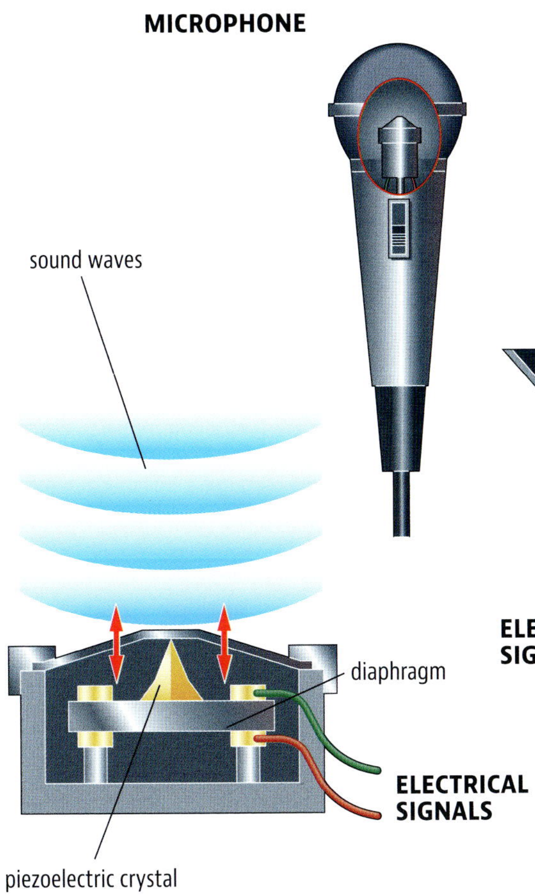

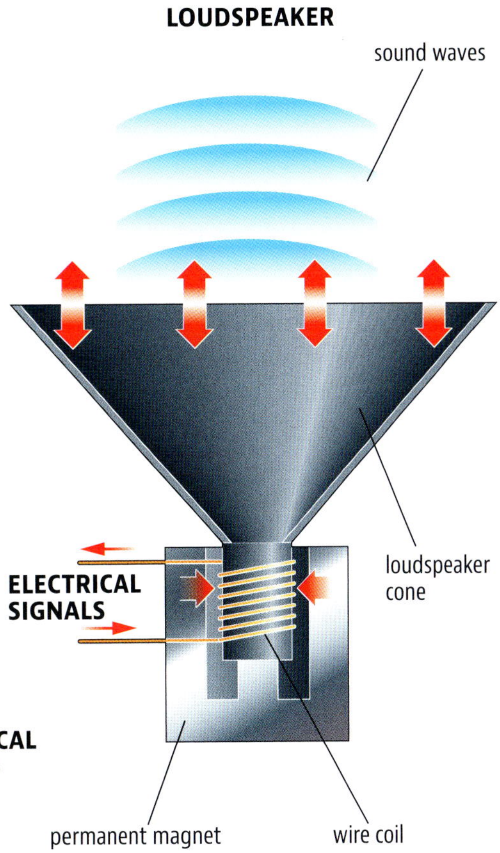

of electromagnetism by British physicist James Clerk Maxwell (1831–1879).

Electric Engines

Until the start of the 19th century electricity was largely a matter of scientific curiosity; it had little practical use. But the principles of electromagnetism were to change all that. In the 1820s, Joseph Henry in the United States and British electrical engineer William Sturgeon (1783–1850) produced working electromagnets (temporary magnets produced by electric currents). These were made from coils of wire wrapped around iron bars, which were connected to batteries. By 1832, William Sturgeon had produced a practical electric motor by making a coil of wire rotate in the magnetic field created by large, permanent

Electric motors can be built to any size. This one can power a car, and runs off a large alternating current (AC). It does the same job as a small direct-current (DC) motor driving an electric toothbrush.

SOCIETY AND INVENTIONS

Early Electric Machines

Simple electric appliances have transformed society, taking much of the hard work out of everyday life. Some of these machines seem so modern that it is hard to believe many of them have been around for more than a hundred years. Electric hot pots were displayed at the Chicago World Fair in 1893, but because the original models used bare-wire filaments, they were inefficient and dangerous. The modern pot with its enclosed filament was not developed until 1921.

American engineer Alva J. Fisher (1862–1947) invented the electric washing machine in 1907; electric dishwashers followed five years later. The General Electric Company sold electric toasters in the United States from 1913, although they only toasted one side of the bread at a time. The pop-up toaster was invented 14 years later by American mechanic Charles Strite (1878–1956). Colonel Jacob Shick (1877–1937) patented the electric razor in 1928.

In 1935, Adolph Rickenbacher (1886–1976) placed a microphone on the back of a guitar to produce his Electro Vibrola Spanish Guitar, creating the first electromagnetic pick-up, which converts sound waves into electrical waves.

The design of the electric food mixer has changed little over the last century.

ELECTRIC MOTOR

An electric motor works like an electricity generator, only in reverse, converting electrical energy into mechanical rotation. It relies on the principles of induction, in which running a current through a metal wire creates a magnetic field around it. The temporary electromagnetic field interacts with two permanent magnets in the motor, making the wire move in response.

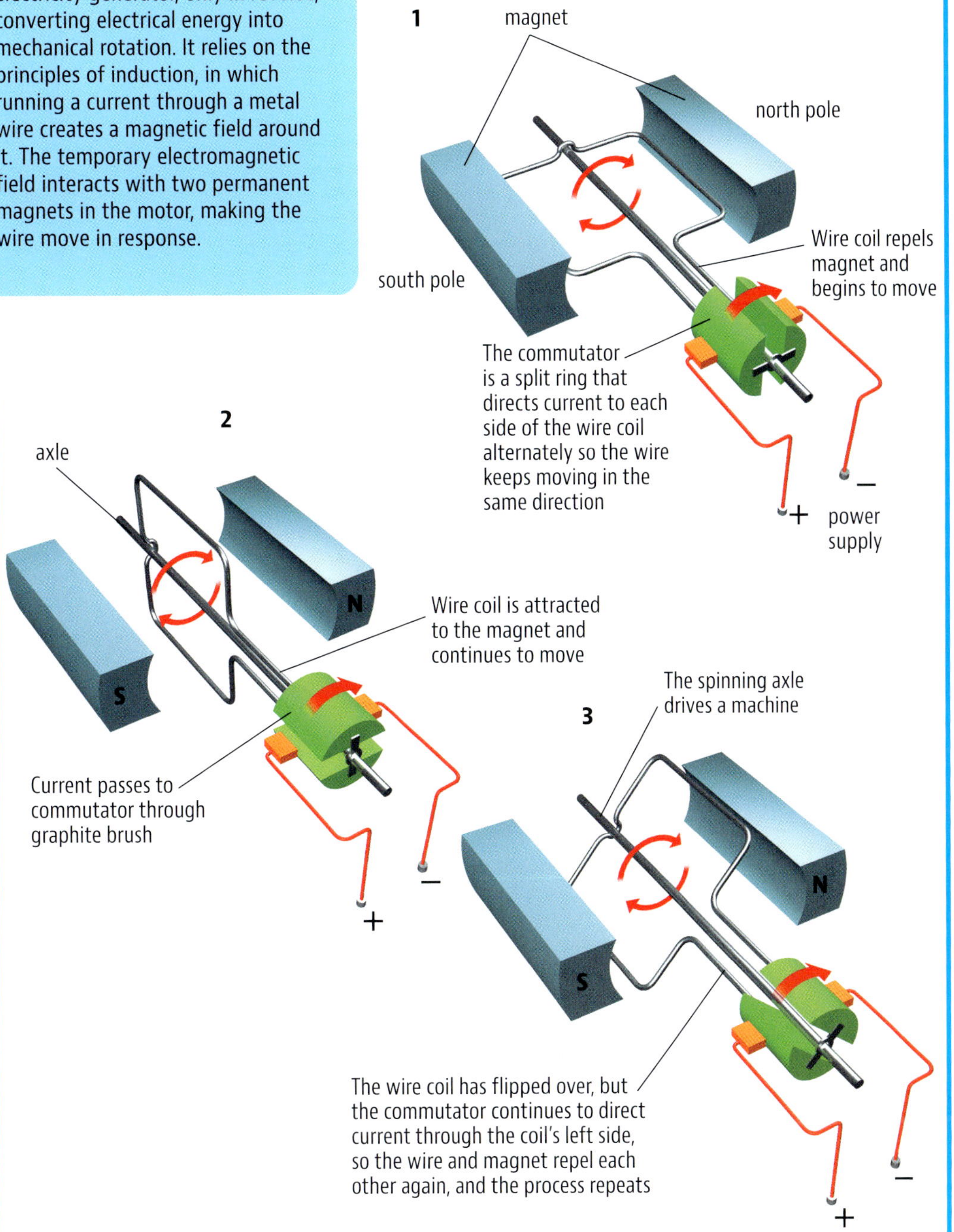

Power plants have giant cooling towers, where hot water is left to steam until it is cold enough to be reused.

FACTS AND FIGURES

- Thomas Edison's Pearl Street power plant of the 1890s in New York used six generators to produce a then enormous 671,130 watts of electricity. When it opened, it was powering 1,300 electric lamps for 59 customers.
- The largest hydroelectric power plant in the world is the Three Gorges Dam Plant in Hubei Province, China. The plant consists of 32 main turbines linked to two smaller generators, which can generate an incredible 22,500 megawatts (22,500,000,000 watts) of power. It supplies electricity across an enormous area, to hundreds of millions of people from Shanghai and Guangdong to Changzhou.

magnets. Sturgeon invented a device called a commutator, which reverses the electric current every half-turn of the motor and so keeps it rotating in the same direction. This invention has been used in most electric motors ever since.

A similar electromagnetic engine was patented in the United States in 1837 by blacksmith-turned-inventor Thomas Davenport (1802–1851). It was later developed by American inventor W.H. Taylor. Taylor's machine looked like a small wooden water wheel mounted on trestles. Seven coils of iron were fastened to the wheel and four electromagnets were attached to the frame. The wheel rotated when the current was switched on.

The first commercial electric motors were built in 1873 by Belgian engineer Zénobe T. Gramme (1826–1901), who equipped his factory with electric motors. Just as electricity can produce magnetic effects and rotate an electric motor, so magnetism can be used to generate electricity. Practical generators were developed from Michael Faraday's early research from the 1830s onward. Originally called magnetos, they were made on a commercial scale by Gramme from 1870 onward. He was the first person to produce a reliable current.

The development of the arc light in the 1840s and filament lamps in 1879 led to an

Overhead electric cables are made from aluminum, which has a low resistance and is also very light.

POWER GRID

Electricity has the great advantage of being a transportable energy. Electricity is generated at the power plant (**1**). A step-up transformer (**2**) boosts the voltage so the current flows efficiently through the grid (**3**). The voltage is reduced at a substation (**4**), and a step-down transformer lowers the voltage further (**5**).

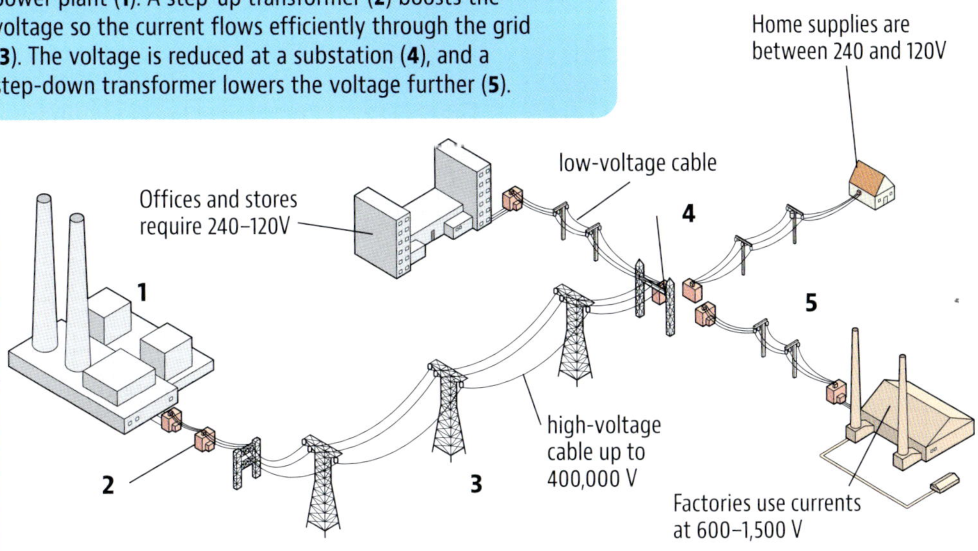

A circuit board connects electronic components that between them can combine in millions of different ways. Current runs between them along metal connections printed on the base board.

enormous interest in generating electric power on a large scale. In 1879, electric light bulbs began to be used for street lighting, while the production of electric motors, heating filaments, and electrical wires to enable long-range communication (by telephone, for example) confirmed that the age of electricity was at hand.

This all became even more important when ways were found to generate electric power reliably on a large scale and then transmit it to wherever it was needed. The first of these "power stations" were at Holborn Viaduc in London, England, and Pearl Street in New York, both constructed in 1882 by Thomas Edison (1847–1931).

In the Modern World

The great revolution in the way electricity was harnessed in the 20th century was the miniaturization of electrical components. Very small machines could be created that were driven by electricity. The discovery of the electron, the electrically-charged particle that is part of all atoms, in 1897 led to the development of various new devices such as vacuum tubes and cathode ray tubes. Such devices were the basis of the invention of television in the 1920s. The ability to control electricity was then revolutionized by the invention of the transistor in 1947. This was followed by the development of semiconductors, small components (nowadays mostly made from silicon) that enabled very precise control of complex electronic circuits.

Today, many people carry a sophisticated computer almost everywhere they go, in the form of a smartphone.

Today's computers run on a binary system of electrical signals that are either on (1) or off (0). Scientists are investigating computers that would use the power of atomic particles to operate. These quantum computers would be much faster and more powerful than digital computers. Quantum computers are not yet practical but might one day be working alongside our classical computers.

INSIDE ELECTRONIC EQUIPMENT

COMPONENT	FUNCTION	EXAMPLES	
Capacitor 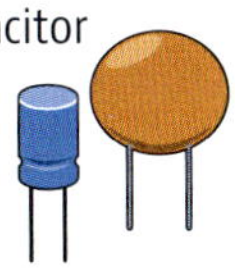	A device for storing electric charge. It is made from two conducting plates (sheets of metal) separated by an insulating material called a dielectric.	A radio tuning dial. Turning the dial alters the charge stored and changes the station the radio picks up.	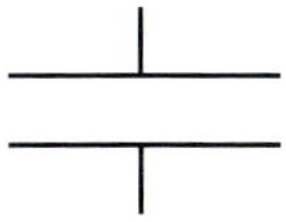
Diode 	An electrical "one-way street," current can flow through it in one direction only. Diodes can be used to convert alternating current into direct current.	A light-emitting diode (LED) gives off light when current flows through it and is used in many electronic devices.	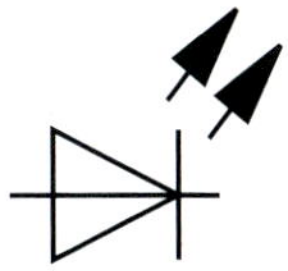
Resistor 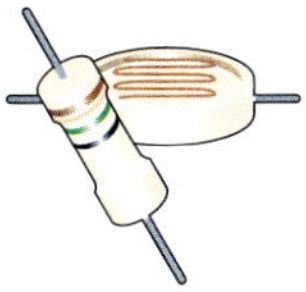	Controls the flow of current through an electric circuit. Resistance (measured in units called ohms) relates to the voltage needed to make a current of a particular size flow through a particular component.	Volume controls are often variable resistors known as potentiometers. Turning the dial adjusts the resistance and changes the volume.	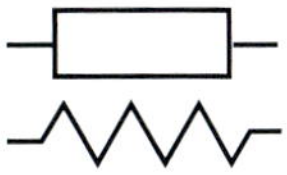
Vacuum tube 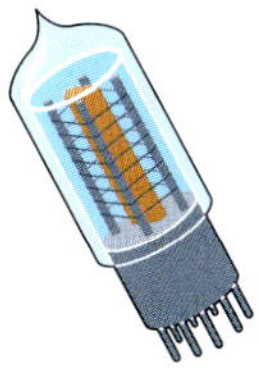	A tube of low-pressure gas that regulates the current flowing through a circuit. Vacuum tubes have three electrodes, one of which is called a grid. By altering the grid voltage, the current flowing through the tube can be increased or decreased.	Vacuum tubes were first used to amplify the signals in early radios. However, the devices consumed a large amount of power and were quite unreliable.	*Vacuum tubes are no longer used and do not have a circuit symbol.*
Transistor 	A transistor is the modern-day equivalent of a vacuum tube. Like a vacuum tube, it has three electrodes and can work as either a current amplifier or a tiny electronic switch. However, unlike a vacuum tube, it is very reliable and uses hardly any power at all.	Transistors can be combined to make simple switches called logic gates. These devices enable computers to store information and follow their programming.	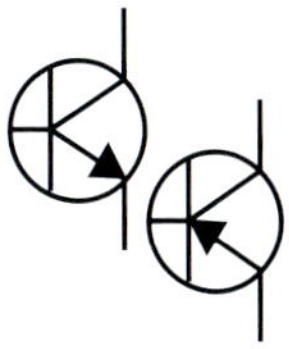
Integrated circuit 	An integrated circuit is a collection of electronic components produced in miniature form on a slice of silicon called a silicon chip or microchip. Silicon chips are the brains behind almost every electronic appliance around today.	The most complex integrated circuits are microprocessors, which may contain millions of transistors and form the core component of a computer.	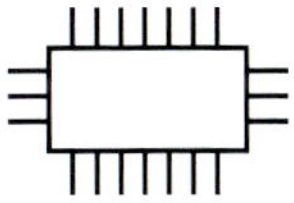

POWERING THE FUTURE

Steam rises from a nuclear power plant, where electricity is generated without burning fossil fuels.

By inventing machines and harnessing the power locked away in fuels, humans have changed the world dramatically, but this has come at a cost to the planet: pollution and climate change. In future, we will need to find more alternative sources of power.

Humans have converted a huge amount of the energy stored in Earth as fossil fuels (coal, oil, and gas) into heat, materials, and transport around the globe. No one could doubt that we have been remarkably successful at harnessing power and energy to fuel the world's economic development. However, another result has been smog-choked cities, polluted oceans, burning rain forests, and global warming—all signs of an energy-hungry world that may be spinning out of control. One of the challenges facing scientists today is to develop energy sources that allow the world to continue to develop economically without destroying the environment. More than 50 years ago many people believed nuclear, or atomic, power could offer a way of generating huge amounts of cheap electricity, and with a minimum of

FACTS AND FIGURES

- Fossil fuels took hundreds of millions of years to form inside Earth's crust. Humans have used around 80 percent of this energy reserve in the 60 years between 1960 and 2020 alone. Oil reserves may only last another 50 years.
- In 2024, figures showed that the United States consumed 16 percent of the world's energy, while only four percent of the world's people lived there.
- Each citizen in the United States uses five times as much energy as someone in Brazil and 30 times as much as a person living in Bangladesh.

pollution. Today, it is uncertain what role nuclear energy may play in the future.

Nuclear Power

It was not until the 1940s that the power of atoms was first successfully harnessed. In 1940, the U.S. government sponsored the Manhattan Project to develop atomic weapons for use in World War II (1939–1945). Leading nuclear physicist Enrico Fermi (1901–1954), who had fled to the United States from Italy, powered up the first nuclear reactor in December 1942. This formed the basis of the atomic bombs that were dropped on the Japanese cities of Hiroshima and Nagasaki toward the end of the war. The world's first commercial nuclear power plant did not begin operating until October 1956. Named Calder Hall, it was opened in the remote county of Cumbria in northwest England. The site is now named Sellafield and remains one of the world's major nuclear plants.

DISCOVERING RADIOACTIVITY

The history of nuclear energy dates from 1896, when French physicists Antoine-Henri Becquerel (1852–1908) and husband and wife Marie (1867–1934) and Pierre Curie (1859–1906) first explored radioactivity. They found a process in which certain atoms broke up, releasing energy. The trio shared the 1903 Nobel Prize in Physics for their work.

The Curies discovered two radioactive elements, radium and polonium, named for Marie's home country of Poland.

SOCIETY AND INVENTIONS

The Invisible Power Station

Influential American physicist Amory B. Lovins (b. 1947) believes the best way to generate power is to save energy. He argues that around 50 percent of the energy we use can be saved with no loss of living standards. He uses the term "negawatts" to describe the power that can be effectively generated by using energy-efficient systems.

Utilities companies have given away millions of low-energy lamps, saving power at a fraction of the cost of building a new power plant. In 2008, a British gas company gave away 52 million energy-saving light bulbs.

This futuristic home is designed to save energy and is also constructed from non-polluting materials that will not harm the environment.

Wind and Water

Fossil fuels produce pollution and global warming and seem to be running out at an alarming rate. Many people believe the main alternative, nuclear power, is neither safe nor cost-effective. The search is on for renewable energy to produce power without pollution. The world produced about 80,000 times more electricity by wind power in 2023 as it did in 1983, while the biggest turbines can produce 100 times more power than they did then.

In the mid-19th century, American hydraulic engineer James B. Francis (1815–1892) had developed a turbine with curved blades at

RADIATION DANGER

Around 400 nuclear power plants operate around the world, but the technology has not been without its problems. It has been expensive to operate safely, disposing of nuclear waste is extremely difficult, with many environmental concerns, and several nuclear accidents have led to widespread public alarm.

In 1979, at Three Mile Island in Pennsylvania, some 200,000 people were evacuated following an accident at the nuclear plant. An explosion at Chernobyl, Ukraine, in 1986, threw a cloud of radioactive dust, or fallout, across much of Europe. Over 100,000 people received a dangerous dose of radiation. In 2011, around 18,500 people died as a result of the combined effect of an earthquake and tsunami that caused a nuclear disaster in Fukushima, Japan.

Nuclear fission is controlled inside a plant's nuclear reactor. This is housed inside an extremely strong concrete dome that can withstand explosions.

SPLITTING ATOMS

Two types of nuclear reaction are used to generate energy. The first, called fission, is used in nuclear power stations and involves splitting atoms. Some elements, such as uranium, exist in a stable form, uranium-235, and an unstable form, uranium-236. When an atom of uranium-235 is hit by a particle called a neutron (**1**) it changes into unstable uranium-236. This splits into two smaller atoms and gives off more neutrons and large amounts of energy (**2**). The newly released neutrons convert more uranium-235 into uranium-236 (**3**), starting a self-sustaining process called a chain reaction (**4**).

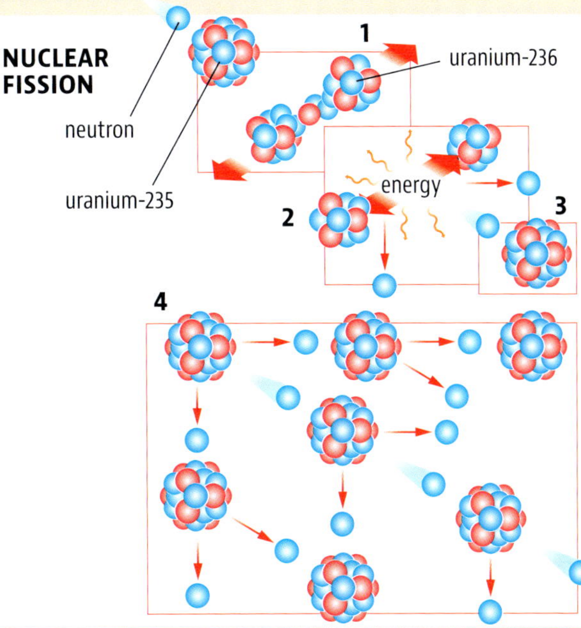

JOINING ATOMS

Another nuclear reaction is called fusion, which produces energy by joining lighter atoms together to make heavier ones. Two heavy forms of hydrogen, called deuterium and tritium, are suspended in a magnetic field (**1**) and smashed together at high speed (**2**). Their nuclei fuse (**3**) to produce a neutron, a helium atom, and energy (**4**). Nuclear fusion produces more energy than fission and is used in hydrogen bombs. However, it may be some decades before this reaction can be controlled and used to produce energy for usein power stations.

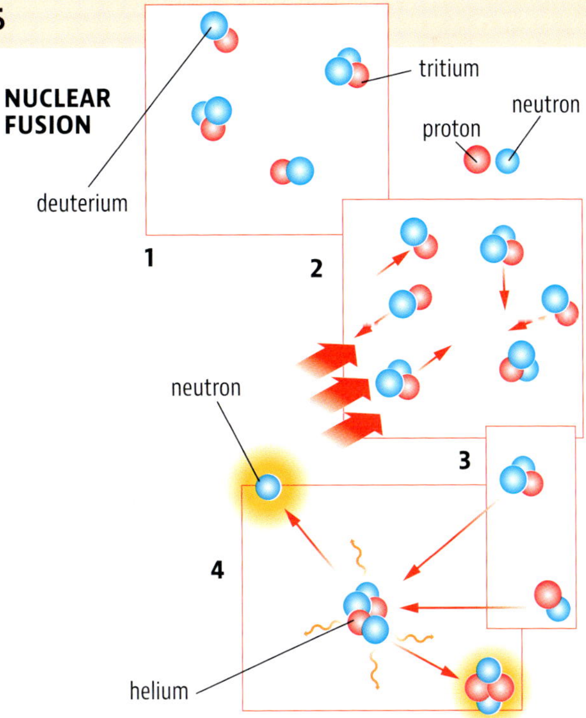

just the right angle to extract maximum power from the water rushing past it. The Pelton wheel, designed in 1889, had small cups mounted all the way around its circumference that caught the water and turned the wheel.

Wind turbines are clustered together in "wind farms" out to sea, or on hilltops, where winds are strongest.

REINVENTING WINDMILLS

Windmills have been used for 1,400 years, but there were few innovations until the 20th century. Efficient wind turbines have been developed that use the power of the wind to turn a generator and produce electricity. In 1931, French engineer G. Darrieus (1888–1979) designed a turbine that turned on a vertical axis so that it could catch the wind from any direction. A prototype of the modern turbine, using a set of blades shaped like an aircraft propeller, was tested in the 1940s but abandoned when a blade broke and flew through the air. The first successful propeller turbine was installed near Sandusky, Ohio, in 1976. Wind power now produces more energy than any other renewable source. In 2023, ten percent of energy in the United States was provided by wind power.

Water Power

Hydroelectric power plants are effectively water wheels operating on a vast scale. They consist of a number of turbines placed in the path of a fast-flowing river, a river that has been dammed, or a sea dam built across a tidal estuary. The first hydroelectric power plant was built at Appleton, Wisconsin, in 1882. In 1936, the giant Hoover Dam was completed on the Arizona-Nevada border and contains 17 Francis turbines.

Despite their obvious attractions, wind and water power have their problems. Hydroelectric dams can be environmentally destructive. The world's largest, the Three Gorges Dam across the Yangtze River in China, flooded an area 410 miles (660 km) long. It took four years for the reservoir to fill after the dam was completed in 2006, and 1.3 million people were forced out of their homes.

HYDROELECTRIC POWER PLANT

Water power is one of the most useful renewable energy sources. There are several efficient turbine designs, such as the Pelton wheel, the Francis turbine, and the Kaplan turbine (shown here). The height of water stored behind a dam is called the head. The head and water flow rate determine the choice of turbine for a particular dam. The Kaplan turbine is used for low heads and high flow rates. Guide vanes and a control gate regulate the amount of water reaching the turbine, which is completely submerged in a vertical shaft. The water flow turns the turbine wheel, and the motion produced is transmitted via an axle to a generator, which converts it into electric power.

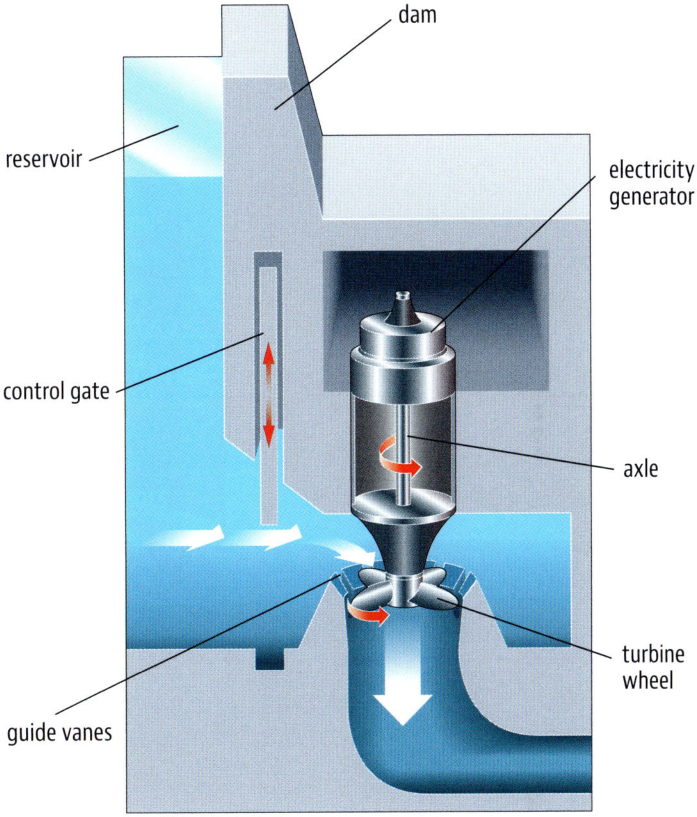

As for wind power, many places are just not windy enough to generate enough power. In 2023 in the United States, just five states between them generate around 60 percent of the entire country's wind-generating capacity.

Energy from the Sun

Most of the world's energy sources, including wind and water power and all fossil fuels, come indirectly from the Sun. Another way of generating power is to use the energy from

The water used in power generation is channeled through a dam and then powers turbines. When water levels are too high, the extra water is released over the top.

SOLAR CELLS

In 1887, German physicist Heinrich Rudolf Hertz (1857–1894) discovered that light could be used to generate electricity. Called the photoelectric effect, this phenomenon led to the development of the first solar cell in 1889. However, it was not until silicon solar cells were developed by the American engineer Russell Ohl in (1898–1987) in 1941 that the generation of electric power from sunlight became practical.

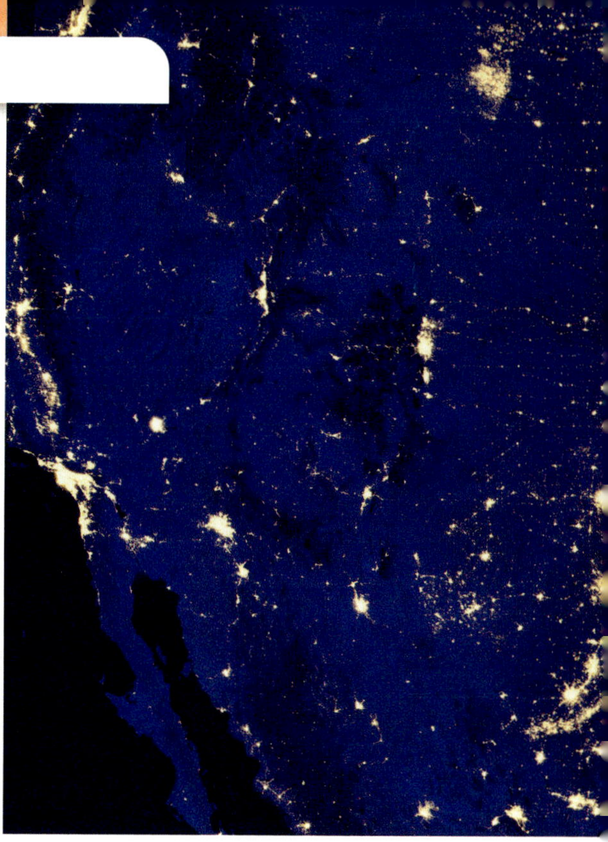

Thermal solar power plants use mirrors that direct the Sun's heat and light onto water pipes. The water boils, creating a flow of steam that drives a generator.

The true impact of electric power is shown when the surface of Earth is seen at night from space. Here, electric lights illuminate huge areas of North America.

the Sun directly to heat water or buildings, or to convert it into electricity with arrays of mirrors or solar cells. People first used solar electricity in the United States in 1954. By 2020, solar was providing 3 percent of all US electricity. This is predicted to increase to 20 percent by 2050.

Power for the Future

Renewable energy has become more important over the past decade. It only produced a tiny fraction of the world's energy in the year 2000, but it is anticipated that it will be responsible for 50 percent of the world's energy by 2050. As the world's population increases rapidly, never has the role of engineers and scientists been more important for the future of the planet.

GROWING NEW FUELS

One solution to the world's falling supplies of oil could be to run vehicles on other liquid fuels. Two alternative fuels, gasohol and biodiesel, have already become popular. Gasohol is a mixture of about 90 percent gasoline and 10 percent ethanol (this chemical is also in alcoholic drinks) that is designed to make the gasoline last longer. Ethanol is usually made from agricultural waste products or crops, such as sugar beet or cane, so is economical and renewable. However, there are concerns that the pollution released and the energy used up in its manufacture outweigh gasohol's benefits.

Unlike gasohol, biodiesel contains no alcohol. Instead, it is made by processing waste vegetable oils and animal fats. This processing produces chemicals called methyl esters that can be burned in modified diesel engines. Biodiesel is a very environmentally friendly fuel: It is nontoxic, breaks down naturally in the environment, and can produce less harmful air pollution when burned.

A large proportion of the sugar cane crop grown in countries such as Brazil is now used to make fuel, not food.

TIMELINE

1,800,000 BCE The earliest tools yet discovered are made around this time in the Olduvai Gorge, Tanzania.

600,000–500,000 BCE Humans discover how to make fire.

5000 BCE Mesopotamians exploit wind power using sails.

3500 BCE Wheels are added to sleds to make the first carts.

2900 BCE The earliest dam is built on the Nile River in Egypt.

600 CE Large windmills are built in Persia.

1712 Thomas Newcomen develops a steam engine for removing water from mines.

1782 James Watt invents the double-acting steam engine, using two cylinders.

1787 Oliver Evans builds the first fully automatic grain mill.

1800 Alessandro Volta invents the voltaic pile, the first electric battery.

1801 Joseph-Marie Jacquard devises an automatic loom. Richard Trevithick builds a steam-powered carriage.

1820s Joseph Henry and William Sturgeon produce the first working electromagnets.

1821 Charles Babbage constructs the Difference Engine, the ancestor of modern computers.

1859 The world's first oil well is sunk at Titusville, Pennsylvania, beginning the oil industry.

1867 Nikolaus August Otto makes the first working internal-combustion engine that uses the four-stroke cycle.

1870s The assembly line—a mechanized production technique—first appears in the meat-packing factories of Chicago.

1873 The first commercial electric motors are built by Zénobe T. Gramme.

1878 Joseph Swan invents the light bulb, and goes into partnership with Thomas Edison to manufacture them.

1879 Inventors Werner von Siemens and Johann Georg Malske unveil their electric locomotive.

1882 The first hydroelectric power plant is built at Appleton, Wisconsin. Coal-fueled power plants are built at Holborn Viaduct in London and at Pearl Street in New York City to generate electricity.

1887 Heinrich Rudolf Hertz finds that light can be used to generate electricity, a phenomenon called the photoelectric effect.

1892 Rudolf Diesel patents the diesel engine.

1912 The first diesel locomotive is built.

1942 Enrico Fermi sets up the first self-sustaining nuclear chain reaction in Chicago.

1945 The first atomic bombs are dropped by the United States on the Japanese cities of Hiroshima and Nagasaki.

1947 Scientists at Bell Laboratories invent a type of miniature electronic switch called a transistor.

1952 The first hydrogen bomb, which uses nuclear fusion, is tested.

1956 The first nuclear power plant opens at Calder Hall, England.

1959 Scientists working at Texas Instruments develop the integrated circuit.

1962 The first industrial robot is sold to General Motors.

1986 Explosion in a nuclear reactor at Chernobyl, Ukraine, creates the world's worst nuclear accident.

1990s Lead gasoline is phased out to reduce pollution.

1997 The Toyota Prius, the first commercially available hybrid car is released.

2011 The Fukushima nuclear plant in Japan goes into meltdown after an earthquake.

2024 The world's biggest offshore wind farm is completed off the Netherlands. Its 139 turbines produce 1.5 GW (1500 MW) of power.

GLOSSARY

alternating current A type of electric current in which the direction of flow changes at regular intervals.

amplify To make an electronic signal larger or stronger. Amplifiers are devices that amplify signals and are widely used in electronic equipment such as televisions, computers, and radios.

ancient Greece A civilization that existed on the mainland and islands of modern-day Greece and Turkey between 2000 and 300 B.C.

anode A positively charged electrode, which usually attracts negatively charged particles.

atom The smallest unit of a chemical element that can take part in a chemical reaction. Atoms are composed of a central nucleus made up of protons and neutrons surrounded by shells of electrons.

cathode A negatively charged electrode, which usually attracts positively charged particles.

compression The act of being made more compact by the application of pressure.

conductor Any material that can easily transmit heat or electricity.

direct current An electric current that flows in a constant direction.

electrode A conductor that carries electric current into or away from the other conducting substances in an electric circuit. Electrodes are often used to make electrical contacts with nonmetals.

electromagnet An iron core surrounded by a coil of wire that temporarily generates a magnetic field when an electric current flows through the wire. This phenomenon is called electromagnetism.

electromagnetic radiation Waves made up of electric and magnetic fields. Electromagnetic waves travel at the speed of light, but the characteristics of a particular wave are determined by its frequency. Radio waves and microwaves are low-frequency electromagnetic waves, while infrared, visible light, ultraviolet, and X-rays are produced as the frequency increases.

electrons Negatively charged particles that are basic constituents of atoms. Electrons are involved with many phenomena such as the flow of electric currents through circuits.

fossil fuels The fuels made from the remains of trees and other living things that have been buried underground for millions of years. Fossil fuels include natural gas, gasoline, and coal.

global warming The gradual warming of Earth due to heat from the Sun being trapped by certain gases building up in the atmosphere. The most important of these gases is carbon dioxide, and its levels are rising largely because of human activities such as the burning of coal and oil.

induction The phenomenon in which a magnetic field creates an electric current in a wire moving through it.

Industrial Revolution A great change in social and economic organization brought about by the replacement of hand tools by machines and power tools, and the development of large-scale industrial production methods. The Industrial Revolution started in England around 1750 and spread to the rest of Europe and the United States.

insulator A material that does not allow heat or electricity to flow through it easily.

ion A charged particle created when a neutral atom or molecule loses or gains an electron.

magnetism All phenomena associated with magnets and magnetic fields. Magnetic fields are regions around magnets in which a force acts on any magnet or electric charge present.

magnets Any material capable of generating a magnetic field. See also magnetism.

mechanical advantage A measure of how much a machine alters the force applied to it, either reducing or amplifying it.

medieval Of or from the Middle Ages.

Middle Ages A period of European history that ran from around A.D. 500 to c. 1450.

neutron A component of an atom's nucleus. Neutrons do not have an electric charge.

Persia Modern-day Iran.

petroleum A mixture of hydrocarbon oils and tars that occurs naturally underground. Also known as crude oil. The word *petroleum* comes from the Greek and Latin for "rock oil."

piezoelectric Able to produce electricity when subjected to a force. The piezoelectric effect is found in certain types of crystal.

protons Basic components of atoms. Protons are located in the nucleus and have a positive electric charge.

radiation The act of giving off radioactive particles, heat, or electromagnetic waves.

radioactivity The disintegration of *atomic nuclei*—the cores of atoms—accompanied by the giving off of particles or electromagnetic waves.

resistance A characteristic of a substance that relates to how easily electric currents flow through it. Materials with a high resistance to electric currents are insulators. Thosel with a low resistance are conductors.

solar panels Collections of solar cells that are used to convert sunlight into electrical energy.

superconductor A material that can conduct electricity without any resistance. The few materials that exhibit superconductivity only do so at extremely low temperatures.

thermal Relating to heat and the way it moves through a material and is transferred from one substance to another.

torque The turning effect of a force.

transformer An electrical device for changing the voltage of an electric current.

turbine A machine made up of a set of blades mounted on a central shaft. A moving fluid, such as steam or air, makes the assembly rotate. Turbines are often used to drive generators.

voltage The force that pushes a current through a substance. Currents require only a small voltage to flow through conductors, but only a very high voltage can push a current through insulators.

FURTHER RESEARCH

Books

Machines Through the Ages: From Furnaces to Factories by Michael Wood and Mary Wood. Minneapolis MN: Lerner Publishing, 2024.

Power Stations and Electricity by Paul Mason. New York: Rosen Publishing, 2024.

Understanding Basic Electricity by Irwin Math. Westfield NJ: Artisan Ideas, 2023.

Web Sites

Energy Kids
www.eia.gov/kids/

How an Electric Motor Works
www.explainthatstuff.com/electricmotors.html

How Nuclear Fusion Works
science.howstuffworks.com/fusion-reactor.htm

INDEX